THE KITTIWAKE

THE KITTIWAKE

John C. Coulson

Illustrated by
Robert Greenhalf

T & AD POYSER

London

Published 2011 by T & AD Poyser, an imprint of Bloomsbury Publishing Plc,
49–51 Bedford Square, London WC1B 3DP

ISBN (print) 978-1-4081-0966-3
ISBN (e-pub) 978-1-4081-5233-1
ISBN (e-pdf) 978-1-4081-5234-8

A CIP catalogue record for this book is available from the British Library

This book is produced using paper that is made from wood grown in managed sustainable forests. It is
natural, renewable and recyclable. The logging and manufacturing processes conform to the environmental
regulations of the country of origin.

Commissioning Editor: Jim Martin

Design by Mark Heslington Ltd, Scarboroug

Printed in Great Britain by the MPG Book G

10 9 8 7 6 5 4 3 2 1

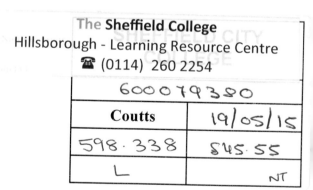
Visit *www.acblack.com/naturalhistory* to find out more about our authors and their
books. You will find extracts, author interviews and our blog, and you can sign up for
newsletters to be the first to hear about our latest releases and special offers.

Contents

For David, Steven and Joey

Acknowledgements

MY thanks go first to Edward White, who shared the study of the kittiwake with me up to 1960. We developed a most productive and effective partnership and I was very sorry that his appreciable contribution had to end when he moved to a permanent post in New Zealand. They also go to Fred Gray, who introduced me to birds and their ecology.

My unfailing thanks go to Andy Hodges, Jeff Brazendale, Julie Porter, Callum Thomas, Ron Wooller, Jackie Fairweather and John Chardine, all of whom made studies of the marked kittiwakes at the North Shields warehouse, and also to Fiona Dixon, who used the neighbouring colonies at Marsden for her studies, and to Tom Pearson, who studied the food and feeding of kittiwakes and other seabirds on the Farne Islands. Etienne Danchin spent a summer in north-east England and Scotland, and helped with work at North Shields and elsewhere.

Professor J. B. Cragg gave much encouragement and advice during the early stages of these studies and without his support they may never have been more than just casual observations.

James Fisher kindly gave me his detailed archive of information on kittiwake colonies in England up to 1955, collected while making his studies on the Northern Fulmar. He also greatly helped with the preparation and logistics of the first national kittiwake survey in Britain and Ireland which was linked with his fulmar enquiry. This allowed me to gain the help of his many contacts, in addition to the British Trust for Ornithology's network of regional representatives.

Dr Geoffrey Banbury, a lecturer in the Botany Department at Durham University who had considerable experience of the use of radioactive materials, helped me greatly in the development of a continuous method of recording, for weeks and even months, the presence at the nest of male and female kittiwakes. Thanks to his expertise in the use of radioactive materials, reliable and clear results were obtained from the very first attempted use of the method and I am most grateful for his expertise.

I am grateful to those who made large numbers of *Darvic* colour-rings for me once I had developed the methodology. My thanks goes first to Jack Warner for making hundreds of the single-colour rings for this and for other studies. He subsequently received large orders from other researchers and set up what appeared to be a profitable home industry! Later, Eric Henderson and Michael Bone continued the production of colour rings for me and also efficiently made up to two hundred laminated rings each year engraved with

alpha-numeric codes. Their technical support was valuable and their forti-
tude when suffering from sore fingers arising from this work was very much
appreciated. I am also grateful to Alf Bowman for his electrical expertise in
ensuring that equipment used in the studies was regularly overhauled, and
for setting up satisfactory sources of electrical power at North Shields.

I am particularly indebted to John Strowger and also members of the
Durham Bird Club for their interest in the kittiwakes at Marsden and for
collaboration in marking of kittiwakes nesting there in the 1990s, and to
Danny Turner, who has continued to monitor kittiwakes breeding along the
River Tyne and elsewhere in north-east England.

Access to the warehouse at North Shields was facilitated by Smith's Dock
Co. Ltd, Jimmy Nessworthy and Jim Pepper. Access to the Baltic Flour Mill at
Gateshead was given by Rank Hovis Co. Ltd and later by an unknown person
who gave me a set of the keys to the building when it became unoccupied!

Over the years, I have benefited from considerable correspondence with
many people throughout the northern hemisphere with interests in kittiwakes, all too many to list individually, although I have to single out the
particular help from Bob Furness, Colin Carter and Rob Barrett. I much
appreciate the willingness of many people who have supplied information
about kittiwakes in the North Pacific. Verena Gill and Rob Barrett kindly sent
me copies of their theses, and E. C. Murphy, D. G. Roseneau and A. M.
Stringer went to great lengths to copy and send me their original measurements, including weights of Black-legged Kittiwake chicks in the Pacific, from
which I was able to construct growth curves using comparable methods to
those prepared for the smaller sized kittiwakes breeding in north-east
England. I am also grateful to them for their considerable effort in making
available detailed information on studies they made on kittiwakes in the
North Pacific and upon which I have called extensively in this book. I appreciate the update of information on Black-legged Kittiwakes nesting on
Middleton Island in Alaska supplied by Scott Hatch. Many other people have
interests in kittiwakes and I appreciate their sharing their interest and information with me. Christine Custer kindly sent me information on kittiwakes in
North America which I was unable to obtain elsewhere.

The British Trust for Ornithology made available the many ringing recoveries of kittiwakes ringed in the British Isles, and supplied their metal rings as
and when required. I am particularly grateful to Bob Spencer, when in charge
of the ringing scheme, for his interest in the extent of wear of aluminium
rings on seabirds and his efforts in successfully developing 'hard' metal rings
which did indeed last beyond the lifetime of the kittiwake. The Joint Nature
Conservancy Council made available their database on the size, breeding
success and distribution of kittiwake colonies in Britain and Ireland which
proved to be of great value. I am also grateful to what is now the Natural
History Society of Northumbria and also the National Trust for access to the
Farne Islands and the study centre on Inner Farne, and to Grace Hickling,
John Walton and John Steel for information about kittiwakes on these islands.

Visits and access to other kittiwake colonies were made possible through the Isle of May Bird Observatory, the Dunbar Council, International Paints Ltd and successive owners and managers of the Marsden Grotto (a restaurant and public house surrounded by breeding kittiwakes). The Duke of Northumberland, and more recently the Royal Society for the Protection of Birds, gave permission to study seabirds, including kittiwakes, on Coquet Island off the Northumberland coast.

The research students working at the North Shields colonies were supported by grants from both the British and Canadian Research Councils. The Department of Science and Industrial Research, the Science Research Council and the Natural Environmental Research Council funded parts of the kittiwake studies, while the University of Durham supplied facilities throughout most of the study and made a travel grant available to make the first census of kittiwakes on the north-east coast of England and in south-east Scotland.

Many boatmen have conveyed me to and from islands on which kittiwakes nest and I am particularly indebted to David Gray for the many crossings to Coquet Island and to both the Shiel family and Pat Laidler for visits to the Farne Islands.

I owe a great debt to Michael Osborne for his many excellent photographs, the conversion of old colour transparencies to a digital form, and for much valuable photographic advice he provided. Thanks also to D. V. Kelly for making photographic records of the colonies at Marsden and elsewhere.

I much appreciate the awards of the Godman-Salvin medal from the British Ornithologists' Union and the Robert Cushman Murphy prize from the Waterbird Society as a recognition of the importance of long-term studies on the Black-legged Kittiwake.

Above all, I am greatly appreciative of my wife, Becky, for her consistent support, tolerance and help of all kinds at all times, and I am very much indebted to her for carefully reading and improving the text.

Preface

THIS study on the kittiwake started as a spare-time investigation when my co-researcher Edward White and I were completing our first degree courses in Zoology at Durham University. In the early 1950s, the prospects of being employed as professional ornithologists were poor, with perhaps no more than a handful of people so employed at that time in Britain. How things have changed in 60 years! As a consequence of the poor employment prospects, we both decided to study the ecology of insects in the British uplands, and not birds, for our Ph.D. studies. I studied the many species of daddy-long-legs or craneflies (Tipulidae) in the uplands, but my other interest was happily linked to this by finding a Russian reference to a kittiwake feeding on cranefly larvae!

Edward White and I started our first 'research' by measuring the flight speeds of seabirds between two fixed points on the coast of north-east England in 1952, and during a period when passing seabirds were few, we spent a half hour or so looking at breeding Black-legged Kittiwakes in several colonies nearby at Marsden. This was in late June and we noticed that the pairs in some colonies still had eggs, but no young had hatched, while in others an appreciable proportion of the nests contained chicks and some were already feathered. These differences raised our curiosity and led us to read Rev. E.A. Armstrong's book *Bird Display and Behaviour* where he recounted some of the then current ideas about differences in the timing of breeding in birds in general and the possible impact of social behaviour in colonial species.

In the following year, we started keeping regular, frequent and more detailed records on these and other colonies and so started our research on kittiwakes, which we thought could be completed in two or three years, but for me was going to last a lifetime. In starting this spare-time research, we received much encouragement and interest from Professor J. B. Cragg, then head of the Department of Zoology at Durham. We also benefited from annual visits to the Edward Grey Institute's student conference in Oxford each January, where we made many valuable contacts and had many discussions, particularly with David Lack, Niko Tinbergen, Arthur Cain, many young ornithologists who were yet to become famous, and overseas ornithologists on visits to the UK, including Lance Richdale from New Zealand, who had made long-term and pioneer studies on marked penguins and other seabirds. Mike and Ester Cullen, the latter of whom was then working on kittiwake behaviour on the Farne Islands in Northumberland, were frequent correspondents, referring to us as 'fellow rissaphils'.

One possible cause of differences in the time of breeding in birds pointed out by E. A. Armstrong was the age of the individual birds and he reported that others had found that young birds of several species often bred later than older birds. This suggested that we should study individual kittiwakes of known age, but to do this we needed access to a colony where we could mark and observe individual birds. In 1953, we gained access to a new kittiwake colony I had discovered a few years earlier on a riverside building at North Shields, in north-east England, while crossing the river on a ferry to an inter-school tennis match.

In 1956, I became a school teacher in Durham and in 1957, Edward White moved to a teaching post at Fourah Bay College in Freetown, Sierra Leone, but we both managed to continue studying kittiwakes until 1959, he while on annual leave to the UK, and I in evenings, weekends and during school holidays. At that point in time, Edward moved permanently to New Zealand and changed his research to the chemistry of freshwaters. However, I still remain indebted to him for the effective co-operation he introduced into our work.

In 1959, I was awarded a prestigious two-year Fellowship from the Royal Commissioners for the Exhibition of 1851, an award which I could take up at any university within the UK. After much thought and consideration, I decided against taking up the fellowship at Oxford University, then, as now, a major centre for ornithological and entomological research, and remained at Durham University so I could continue my studies on the upland and, for the first time, include investigations on marked kittiwakes as part of my formal research studies. Two years later, I was appointed to a Lectureship in Animal Ecology at Durham University. For the first time, I was able to make long-term plans for research, both in the uplands and on seabirds, and to develop further my strong conviction in the value of long-term studies.

Two years after I became a permanent member of staff at Durham University, Professor Jim Cragg moved to the post of Director of the Merlewood Research Station of the Nature Conservancy. His position at Durham was filled by Professor David Barker, whose interests in zoology were as different from my own as was possible to achieve. As a consequence, I took over the supervision of a group of research students working on the ecology of upland and moorland, who had been supervised by Jim Cragg, and I also continued to develop my own interests in this field, thus continuing two very different fields of research which I had followed as a post-graduate student. These studies resulted in the first major investigations of the vast numbers of invertebrates (well over 1,000 animal species) in uplands associated with the moorlands in England and Scotland, identifying several new species, and investigated their ecology in this harsh habitat. Currently, these studies have achieved greater importance with the realisation that such large areas of peat are major storage areas of carbon, and that they contain many rare species in need of conservation.

There were three adverse events which occurred during the kittiwake studies. The first was an ill-considered amalgamation of the Zoology and

Botany Departments of Durham University into a Department of Biology, a blanket action which occurred throughout universities in the country, except at Oxford and Cambridge. This reduced the teaching of and research into animal behaviour, ornithology and entomology, while it did little to advance ecology because co-operation in that field already existed. At about the same time, there was a major decline in ship building and repair on the River Tyne, resulting in Smith's Dock being closed. Their building, on which the kitti-wakes nested, was bought by Mr J. Pepper, a local businessman. Fortunately, he permitted access to the building and did not make major changes to it until 1990, when the property was sold for conversion to riverside residences. The kittiwakes were excluded from nesting there in 1991, but by this time, data for the whole colony and the individually marked birds had been obtained for 37 consecutive years and over 2,500 breeding attempts by marked individuals had been recorded in detail. The third adverse event was the high mortality of adult kittiwakes during 1998 in north-east England, which resulted in the death of most of the colour-ringed breeding birds.

Investigations of the distribution and the first national census of kittiwakes in Great Britain and Ireland in 1959 (Coulson 1963a) were made in collabo-ration with James Fisher and combined with a census of Northern Fulmars. In addition to many personal contacts, the British Trust for Ornithology (BTO) and their regional representatives throughout Britain greatly helped in organising counts of occupied kittiwake nests at each colony by numerous amateur ornithologists. This gave a new and very different picture of the distribution and abundance of the species compared with general statements included in many texts, shifting the most abundant areas from the west to the east coast of Scotland. Subsequent census work on kittiwakes was combined with counts of other seabird species, and these have become regular events. On several occasions, the British Trust for Ornithology kindly made available the ringing recoveries of kittiwakes marked in Britain under their national scheme.

In the 1950s, little was known about the biology and distribution of kitti-wakes, but in the past thirty years the literature on the species has become extensive. In this account, I have tried to make a comprehensive presentation of the biology of the kittiwake, and I have also concentrated on the effects of colonial breeding and those factors which cause major differences to the biology of individuals in relation to the performance of their neighbours. Even now, little is known of organisation within colonies of birds. This approach has hopefully given a new and extensive insight into colony struc-ture and what is gained (or lost) by colonial breeding, a topic on which there has been much speculation, but often based on rather limited information. I have not attempted to make this book a thorough review of the literature. Rather I have aimed it in two directions, namely towards the biology of the kittiwake throughout the year, and to give a detailed account of the outcomes of an intensive study of individually identifiable kittiwakes within the context of a colony. I have not included a detailed presentation of the non-social behaviour of individual kittiwakes, as a detailed review would have taken up more space than could be justified in this volume, but detailed accounts appear in Volume 3 of the *Birds of the Western Palearctic* (Cramp & Simmons 1983) and the studies by Knud Paludan (1955) and Ester Cullen (1957). Nor have I included critical accounts of theoretical models and other speculation on kittiwake biology which were not well supported by actual data. In recent years, there has been a plethora of theories and speculation on aspects of kittiwake biology and colonial breeding, based on the current fashion in ecology of producing 'models', highly complicated statistical analyses often based on purely theoretical considerations. I have read many of these and some seem valid, but in many other cases I have rejected their conclusions because they are based on invalid assumptions about Black-legged Kittiwakes and the conclusions are equally likely to be but speculation for which there is no (current) way of judging their validity. If the assumptions are not correct, it is unlikely that the theories can stand critical examination. Where appropriate, I have referred to those which appear to me to be valid.

The extensive quantitative information obtained on individually identifiable kittiwakes and involving over 2,500 breeding attempts has necessitated decisions on how to condense and present these extensive quantitative data in a relatively simple and lucid manner. In preparation, I have made both long and detailed tables of information, but showing these would not be appropriate to this presentation nor appreciated by the reader. In many cases, the data have been reduced to simple graphs, with each point based on extensive data and so having a high degree of accuracy, much more than in many other studies. I have always been a strong believer in presenting conclusions based on quantitative data only if they are justified by strict statistical tests. That rigour has been applied to this book, but I have spared the reader the presentation of details of numerous statistical tests used to determine whether the outcomes are likely to be real, rather than readily attributable to chance. Essentially, no conclusions have been drawn unless they were justified by appropriate statistical tests which indicated the likelihood of obtaining such results by chance were less than one in twenty ($P<0.05$) and often very much less. This policy has been made easier by have very large samples of data, which appreciably reduced the risk of obtaining differences between sets of data by chance. In cases where non-significant differences have been obtained, I have simply referred to these in the text as being 'not meaningful', which to those with statistical understanding means 'not statistically significant'.

Finally, I offer two apologies. I have used 'weight' instead of the scientifically correct term 'mass' as a measure of body size of individuals. My only excuses are that the former word is probably more meaningful to the non-scientist and there is logic in thinking that when you weigh an object you are recording its weight! Repetitive use of 'Black-legged Kittiwake' has been avoided by reducing it to 'kittiwake' throughout the text, except where this species is being compared with, or could be confused with, the Red-legged Kittiwake, in which case it has been referred to by its full vernacular name.

CHAPTER 1

Introduction

KITTIWAKES are small, specialised gulls with lifestyles that vary greatly with seasons of the year. Like most other gulls, their plumage is white and grey, but they differ in having completely black wing tips without white 'mirrors' and shorter legs. During the spring and summer, they nest in dense colonies on sea cliffs, where their loud outbursts of calling, 'kittiwaak', are familiar. They spend the remainder of the year spread across the northern oceans and far from land, competing with Northern Fulmars as the most oceanic of seabirds breeding in the northern hemisphere.

There are only two species – the Red-legged Kittiwake *Rissa brevirostris* and the Black-legged Kittiwake *Rissa tridactyla* – and both breed only in the northern hemisphere. Initially, the Black-legged Kittiwake, first described by Linnaeus in 1758, was included in the genus *Larus* alongside other gulls. In 1826, it was moved into the newly created genus *Rissa*, characterised by the species having much reduced hind toes and short legs. Subsequently, in 1853, Bruch described for the first time the Red-legged Kittiwake and named it *Larus brevirostris*, but it was later moved into the genus *Rissa*, and currently remains there along with the Black-legged Kittiwake.

The much reduced hind toe and short legs are the results of adaptations which reflect the birds' ability to nest on narrow ledges, and their common need to walk only very short distances, normally taking only a few steps at a time. Short legs provide them a low centre of gravity, which greatly assists in withstanding the buffeting from strong, fluctuating winds and updrafts while standing on exposed, narrow ledges of cliff faces. Both species show other adaptations that assist with breeding on cliffs, including well developed and markedly curved claws for gripping the nest, and the unique behaviour of the female in crouching, rather than standing, during mating. The behaviour of kittiwake chicks is unusual in that they do not move from the nest to hide from predators, in contrast to the young of most other gulls and terns, and the pale unmarked down does not camouflage or conceal the chicks. Both chicks and adults defecate over the side of the nest and no attempt is made to keep the edge of the nest and its vicinity free from faeces, resulting in the side of the nest and the cliff face below becoming 'whitewashed', which makes the colony obvious from a considerable distance.

The Black-legged Kittiwake derives its vernacular name from the colour of its legs and the mutual greeting ceremony involving repeated calling of 'kitti-waak, kittiwaak', which in some countries has given rise to its vernacular name of kittiwake. In other countries, the common name has been derived from its specific scientific name of *tridactyla,* meaning 'three-toed', and this and the fact that it is a gull has been translated into the appropriate language, for example, Drieteenmeeuw (Dutch), Dreizehnmöwe (German), Gaviota tridac-tila (Spanish) and Mouette tridactyle (French). These names suggest that they were derived from examining museum specimens rather than from observations of the birds in the field. Differing from both of these derivations, the Icelandic name of Rita is very attractive although I have failed to trace its derivation, other than its being a girl's name. In many different regions, kitti-wakes have been given many vernacular names; the most extensive of these occur in Newfoundland and include Ticklelace, Tickleass, Pinyole, Annett and Ladybird (Montevecchi & Tuck 1987).

The Red-legged Kittiwake received its common name also from its leg colour, but otherwise it has characteristics in common with the Black-legged Kittiwake, including its very similar greeting call and breeding sites, but differs in its shorter beak, larger eyes, darker grey plumage and darker underwing. The two species also differ in the shape of the head, with the Red-legged Kittiwake having a steeper forehead, giving it a more rounded profile. In fact, the juveniles of the two species differ more markedly from each other than the adults and should rarely be confused. In addition to sharing the differences in the adults of the size of the eyes and beak length, the young of the Red-legged Kittiwake totally lack the black band on the tip of the tail, and the wings lack the black carpal bar but have more extensive black on the outer primaries and primary coverts. Very occasionally, Black-legged Kittiwakes have red legs, but these are never as intensely red as in the other species.

Although not widely appreciated, the Black-legged Kittiwake is by far the most numerous gull in the world, with about 9,000,000 adults distributed throughout the northern hemisphere (Table 1.1, Figure 1.1) and probably a

TABLE 1.1 *The approximate numbers of breeding pairs of Black-legged Kittiwakes estimated between 1980 and 2010 in regions of the world. Based on Mitchell et al. (2004), with additions and corrections.*

Country or region	Number of pairs
R.t. tridactyla	
Spain	200
Portugal	10
France	4,700
Great Britain and Ireland	410,000
Germany	6,500
Denmark	625
Sweden	30
Norway	600,000
Svalbard and Bear Island	270,000
Russia (western)	141,000
Faeroes	230,000
Iceland	700,000
Greenland	150,000
Canada	250,000
Total	**2,763,000**
R.t. pollicaris	
USA (Alaska)	770,000
Russia (Eastern)	570,000
Kuril Islands	550,000
Total	**1,890,000**
World total	**4,653,000**

similar number of immature individuals. As a result, it greatly exceeds the numbers of the next most abundant species, the Herring Gull. In contrast, the Red-legged Kittiwake has a total population of about 200,000 adults and, perhaps, a further 100,000 immature individuals, and is restricted to only four breeding areas, all in the North Pacific Ocean.

Both species of kittiwakes have sub-arctic and arctic breeding distributions, and throughout most of their ranges they breed in colonies where the temperature of the seawater is cold and never exceeds 10°C, even in summer (Figure 1.2). There is one exception to this. Black-legged Kittiwakes breed on the eastern coast of the North Atlantic, from south Norway to Spain and Portugal, including the countries immediately adjacent to the North Sea, where summer sea temperatures are higher due to the influence of the Gulf Stream. Clearly, the distribution of kittiwakes is not solely determined by temperature.

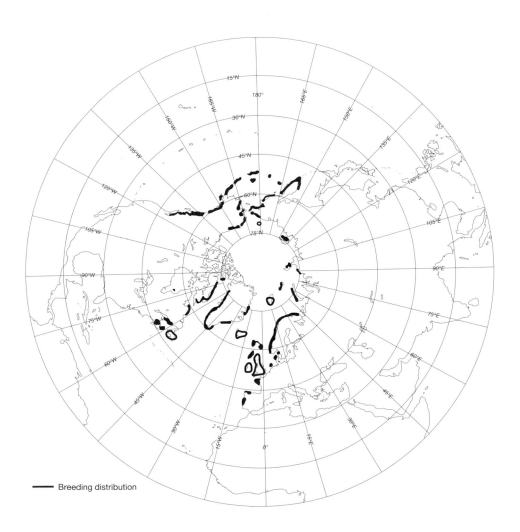

FIG. 1.1 *The worldwide breeding distribution of the Black-legged Kittiwake.*

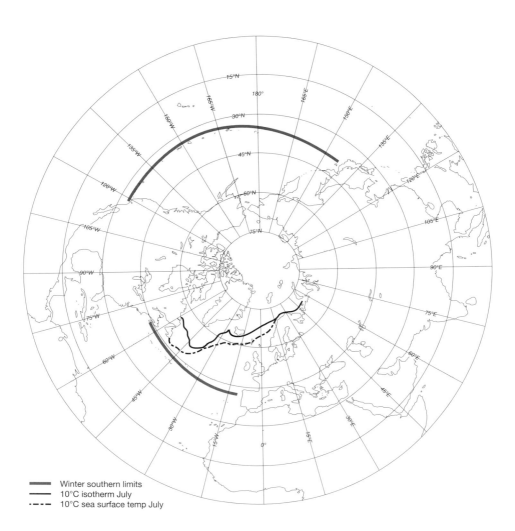

FIG. 1.2 *The southern limits of the Black-legged Kittiwake in January in the North Atlantic Ocean and the suspected limits in the North Pacific Ocean. The winter distribution of the Red-legged Kittiwake is mainly unknown, but is suspected to be over deep water in the North Pacific. The dotted and dashed lines indicate the 10°C July air and sea temperatures, which encompass the breeding area of kittiwakes, except in most of Europe, south of northern Norway.*

THE RED-LEGGED KITTIWAKE

The Red-legged Kittiwake breeds in four locations in the North Pacific, near or on the islands formed by the Aleutian chain. These are Buldir Island, the Bogoslof Islands and the Commander Islands, while to the north of this chain they nest on several of the Pribilof Islands. This rarer of the two kittiwake species is a specialised feeder, adapted for nocturnal or crepuscular feeding, and often forages farther from the colony than the black-legged species, where it exploits squid and Northern Lampfish. Remarkably, the immature (first year) plumages of the two species differ more than those of the adults, with the Red-legged Kittiwake lacking the black tail band, having a much reduced black collar and black carpal bar.

Within its very restricted distribution, the Red-legged Kittiwake invariably nests in mixed colonies with the more numerous Black-legged Kittiwake, which might be expected to result in competition for sites between the species as the sites both select are similar. Mixed pairs have never been recorded.

THE BLACK-LEGGED KITTIWAKE

In future I will often refer to the Black-legged Kittiwake simply as 'kittiwake' as both a means of convenience and to simplify presentation, but I continue to use Red-legged Kittiwake whenever referring to this species. Because it is so much more numerous than the Red-legged Kittiwake, the biology and behaviour of the Black-legged Kittiwake is better understood.

The Black-legged Kittiwake occurs in both the North Atlantic and North Pacific Oceans, where it is represented by the subspecies *tridactyla* and *pollicaris*, respectively. The two subspecies are similar, with the adults of the latter having a slightly darker mantle and more extensive black tips to the primaries. Other differences between the two subspecies which are often used to justify subspecific status are the greater body size of *pollicaris* and also the higher proportion of individuals with a claw on the very small hind toe, but both of these are probably of little real taxonomic value (see below). During the breeding season, the subspecies *tridactyla* shows a considerable and progressive increase in size (cline) from south to north within the North Atlantic (Sluys 1982 and Figure 1.3).

This tendency for individuals to show a geographical change in size agrees with Bergmann's Rule, which states that individuals within a warm-blooded species tend to be larger as the distribution approaches the poles, and so follows the long-known example of the Atlantic Puffin (Huxley 1942). The selective pressure inducing this cline is probably functionally related to the environmental temperature and heat loss from the body, rather than latitude *per se*. The trend shown in Figure 1.3 would have probably been even better if summer sea temperature had been used instead of latitude, particularly since the data used were from similar latitudes in cold Newfoundland and from the

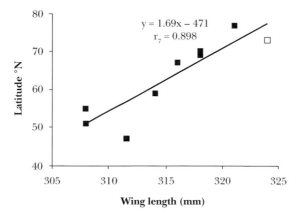

FIG. 1.3 *The relationship between average wing length of kittiwakes and the latitude where they were breeding. The solid squares are data from the North Atlantic and the open square represents data from the North Pacific for the subspecies pollicaris.*

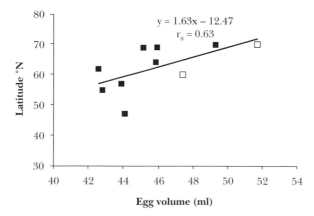

FIG. 1.4 *The relationship between the mean volume of kittiwake eggs and the latitude of the colonies from which they were measured. The solid squares are North Atlantic records and the two open squares are records from the North Pacific for the subspecies* pollicaris.

warmer European coast of the Atlantic influenced by the warm Gulf Stream. It is of considerable importance to note that the position of the Pacific subspecies, *R.t. pollicaris*, in Figure 1.3 falls very close to the trend line of size and latitude for the Atlantic subspecies. As a result, it seems that the reported taxonomic difference in size of individuals of the two subspecies is simply the result of the Pacific subspecies occurring in cold areas and not extending its range as far south as does *R. t. tridactyla* in the North Atlantic. *R. t. brevirostris* has a wing length very similar to the northernmost breeding individuals of *R. t. tridactyla* breeding in similarly cold areas, and this suggests that in both oceans the species is responding similarly to temperature. As a result, the average difference in size between the two subspecies is not of as much taxonomic significance as has been claimed.

Not unexpectedly, kittiwakes also tend to lay larger eggs in more northern areas (Figure 1.4), and again the eggs of the Pacific subspecies fit closely to the appropriate position within this latitude—size trend for *R. t. tridactyla*. It would be interesting to speculate whether the same-sized eggs laid by kittiwakes at the southern end of their distribution have the same yolk and albumin content as those laid further north and whether the chicks arising from them are similar in size and development. I suspect this may not be so.

The proportion of individuals with a claw on the small hind toe is claimed to differ between the two subspecies, and this has been repeated again and again in the literature, but without critical re-examination and consideration. The original reported difference was based on the examination of a small number of specimens in the days before the importance of sample size was fully appreciated. The presence or absence of the claw is problematic to evaluate, since in many individuals it is extremely small and it is often difficult to decide whether it is present or absent or even could be mistaken for a raised scale. At times, the claw appears to have been accidentally lost during preservation or storage as museum skins, since sometimes it is present on one leg and not the other. Further, there is a hint that the proportion of individuals with a claw on the hind toe may vary geographically in *R. t. tridactyla*. If it does vary, like body size, its presence or absence is probably of minimal taxonomic value at the subspecific level, and certainly there needs to be further study before it can be used in justification of the subspecies status. In any case, the taxonomic importance of the claw is reduced because it does not allow identification of individuals, and large samples are probably required to show a meaningful difference in the proportions of birds with claws.

There is usually more extensive black on the tips of the outer primaries in adults of *R. t. pollicaris* than in *R. t. tridactyla*, but there is also more extensive black on the wing-tip of many second- and third-year *R. t. tridactyla* individuals, and so the differences in the extent of the black patterning are not consistent in all age classes, even within each subspecies. Geographical variation in the shade of grey of the wings has not been investigated in the Atlantic population, but the Pacific subspecies is stated to be slightly darker grey, although the extent of this difference remains to be determined using comparable material and applying modern methods of measuring colour shades.

Judging from my own examination of museum material, I believe that the taxonomic differences between the two subspecies *tridactyla* and *pollicaris* have tended to be exaggerated, particularly because the increased similarity of the biometrics of the two at the northern ends of their distributions had not been appreciated. Despite this reservation, and the likelihood that some individual specimens cannot be attributed to a subspecies with complete confidence, the status of the Pacific and Atlantic subspecies is probably justified, and the two represent separate gene pools, further justifying their subspecies status. However, the differences between the subspecies are not large, and there is no justification to even hint that by any criteria they could possibly be considered as distinct species (Chardine 2002).

It remains to be established how far the two subspecies extend along the northern coast of Russia. Fisher & Lockley (1954) showed the point of separation was far to the east, near Wrangel Island (180°E) in the East Siberian Sea and near the Pacific Ocean. James Fisher has said that he has seen evidence in support of this. However, there seems to be a doubt about this since recent photographs of kittiwakes claimed to be taken on Wrangel Island would suggest that they belong to the Pacific subspecies. It still remains to be established how far along the north coast of Siberia the two subspecies penetrate. Do intermediates and mixed pairs occur? At what point along the northern coast of Russia do breeding birds cease to move in winter into the Atlantic Ocean and migrate into the Pacific Ocean, and is there still gene exchange between the two subspecies? There are still many questions to be answered.

The winter distribution of the two subspecies is also separate, with one in the North Atlantic Ocean and the other in the North Pacific Ocean. Only one exception has been recorded, which was a kittiwake ringed as a chick in the Murmansk area of European Russia, well within the distribution of *R. t. tridactyla*, and recovered in winter in Kamchatka, within the Pacific Ocean. It is a long journey from Murmansk to the Pacific Ocean, and the wear and resulting illegibility of numbers on aluminium rings then used on kittiwakes was high. Is this single record spurious, perhaps due to a worn ring being misread? It appears that the ring was not returned at the time of reporting the recovery.

In the Pacific Ocean, the Black-legged Kittiwake has a much more extensive breeding distribution than the Red-legged Kittiwake, with colonies extending from Alaska to Siberia, Kamchatka and the Sea of Okhotsk. On the eastern side of the Atlantic Ocean, the kittiwake nests as far south as France, Spain and Portugal and at the northern end of the distribution breeds on several of the islands north of the Arctic coast of Europe, including Bear Island, Svalbard (Spitsbergen), Jan Mayen Land and Novaya Zemlya. There are large colonies on the coasts of Norway, Iceland, Faeroes and Britain and one small colony in Sweden, one in Denmark and currently a large colony on the island of Heligoland (Germany). On the west side of the Atlantic, kittiwakes started to breed in Nova Scotia about 1970 and they have expanded into the Gulf of St Lawrence, while long-established colonies are frequent in Newfoundland, but only very locally in Labrador and Baffin Island. The breeding distribution is extensive in Greenland, but with many more colonies on the west than on the east coast.

In the United Kingdom, Greenland, Iceland, Faeroes and Norway, the kittiwake is by far the commonest nesting gull species, but this has often been overlooked because they tend to breed in large colonies on high sea cliffs that are difficult to reach and view, while they feed off-shore, well away from most observers. As a result, it is only within recent years that the Black-legged Kittiwake has been recognised as the commonest gull species breeding in the British Isles, and that its worldwide abundance has been appreciated.

TABLE 1.2 *The percentage of time for which the male and female kittiwake faced in different directions at the nest in different stages of the breeding season. By chance, the birds would be expected to spend 33% of the time facing each of the three directions. (Based on Hodges 1974 and time-lapse photography).*

	Early prelaying	Late prelaying	Incubation	Chicks
Male				
Facing cliff	36	35	52	79
Facing sideways	39	41	33	18
Facing outwards	25	24	15	3
Female				
Facing cliff	37	35	46	77
Facing sideways	34	36	35	19
Facing outwards	29	29	19	4

ADAPTATIONS TO CLIFF NESTING IN KITTIWAKES

Both species of kittiwakes show a series of adaptations to nesting on steep cliffs which are mainly restricted to the two species of gull in the genus *Rissa* and are absent in most other gull species. Many of these differences were highlighted by Ester Cullen (1957) in a classic paper and these and others are summarised below.

1. Fighting with other kittiwakes at the nest site occurs in only one way, which involves grasping the beak of the opponent and twisting. The birds lock beaks and often tumble from the nest and spiral towards the sea or ground below before releasing the grip, returning to the nest site and repeating the encounter again and again.
2. The young do not run from the nest when threatened by a predator, but crouch and bury their heads against the rock face.
3. Possession of a black neck-band in chicks and juveniles of *R. tridactyla* appears to reduce aggression at the nest. (This band is much less extensive in Red-legged Kittiwake chicks).
4. Advertising by males on potential nest sites for females is by a 'chocking' display.
5. Pairs are formed only at the nest site.
6. Females do not stand during copulation, but sit on their tarsi and abdomen.
7. Nest material is often collected from sites otherwise not normally visited.
8. Before laying, only the nest site is guarded and defended by the presence of one or both adults during most of the daylight hours. Areas around the nest site are not defended.

9. Nest building is more elaborate than in other gulls, particularly with respect to the use of mud and vegetation to form the base.
10. Prolonged trampling of the nesting material to compact it is frequent and contributes to forming the base of the nest.
11. Typical nests have a deep cup.
12. The young are fed from the throat.
13. The young remain at the nest until fledging and then return there for many days after that point.
14. Unusual objects are thrown over the edge of the nest, including the egg shells after chicks have hatched.
15. Parents do not recognise their own chicks, and will feed any chick on the nest site.
16. In other gulls, the young face in any direction, but in the kittiwake the chicks face towards the cliff face and avoid facing out-over.
17. Both the young and adults have strong and markedly curved claws and toe muscles to aid gripping the nest when landing or retaining position under adverse conditions, such as strong winds.
18. Alarm calls are uttered only when predators (including humans) are very close to the nest and adults often remain on the eggs or with unfledged young even when a predator is close.
19. Droppings are shed over the edge of the nest.
20. The down and plumage of the young are not cryptically marked.
21. Clutch size is lower than in many gulls.

The cliff-facing posture of the chicks is both interesting and involves a complex behaviour. Once, I hand-reared a chick which had fallen from the nest when about three days old. It soon learned to take food from a shallow dish placed on the window ledge where the chick lived. Typically, the chick faced sideways or towards the window and soon recognised when food was being brought to it. However, when I brought the dish towards the bird and then took it on a clockwise curve from near its head towards its tail and then to the other side of the chick, I expected it to follow the dish, as would be done by most animals. However, it did not do so and invariably moved in the opposite (in this case anticlockwise) direction, facing the 'cliff' face the whole time and turning in the opposite direction to which I was moving the dish. This illustrated the very strong motivation in kittiwake chicks not to face away from the cliff face, which would increase the risk of their falling from the nest with lethal consequences.

Andy Hodges (1975) and Edward Burtt (1975) made detailed studies of the direction kittiwakes face when at the nest, and both found that adults and chicks showed a very strong tendency to avoid facing out from the cliff face. A more detailed study of individual adults by Hodges (1974) showed that both sexes behaved in a similar way. During the pre-laying period, there was little selection of the direction to face, although both birds showed a slight avoidance of facing outwards. When together, the pair spend much time facing each other during courtship. During incubation, both birds faced the cliff more than expected by chance. When chicks were in the nest, facing outwards occurred only occasionally, while facing the cliff increased even further, presumably to provide more room for the chicks and keep them towards the back of the nest.

CHAPTER 2

Kittiwake feeding methods and food

MOST animals feed in the medium in which they live and spend most of their time. Fish live and forage in water, foxes on land, earthworms in soil. Swifts spend much time in the air and take flying insects. But most gulls and sea terns are different. They do not feed in the parts of the environment in which they spend most of their lives (land and air), but obtain most food from just below the surface of water. Literally, they can only skim the surface of the sea. In this respect, they are like many kingfishers, but differ in that kingfishers normally exploit shallow waters, whereas gulls tend to use areas where the water and fish therein can be very much deeper. Gulls are only able to catch fish present in the top metre of aquatic environments and for the great part of the time, most of their prey species are in deeper water, beyond their reach and so cannot be exploited. A second category of seabird species, such as auks and cormorants, have overcome this appreciable limitation by being able to swim underwater and are able to reach much greater depths and exploit a much higher proportion of their quarry.

Both species of kittiwakes are able to capture food only within the top metre of the sea surface, and this limitation has several consequences. Firstly, it is unlikely that kittiwakes can ever exploit their food sources to the extent that they cause the numbers and densities of their prey to be appreciably diminished. Secondly, factors that prevent their marine prey entering the surface metre of the sea, such as severe storms, are likely to have an appreciable impact by reducing the feeding success of kittiwakes, because when the sea surface is violently disturbed, the prey species move deeper to avoid the disturbed surface zone. Diving birds can penetrate below the disturbed surface water, but kittiwakes (and terns) cannot. During such periods, Herring Gulls and Great Black-backed Gulls change their feeding habits and forage along shorelines exposed at low tide, in pastures and at landfill sites. Kittiwakes do not show this plasticity, although in some areas they will attend fishing boats while the nets are being hauled. Thirdly, at any time, so few of their prey species are likely to be at the sea surface that kittiwakes can only breed in areas where there are high densities of their prey fish during the breeding season; otherwise they would be unable to obtain adequate numbers of prey near the surface. These limitations might suggest that kittiwakes are maladapted to their marine life and that they employ a high risk breeding strategy, yet the Black-legged Kittiwake is by far the commonest gull in the world. However, large reductions of prey fish populations caused by natural events or over-exploitation by humans are likely to have a severe impact on kittiwakes, and several such situations have already been reported.

In the sea, fish spend most of their time more than a metre below the surface and so are well beyond the reach of kittiwakes. Only some fish species, and often these are restricted to the immature stages, regularly occur in the surface metre of water. What causes fish to visit the top metre of the sea? Some shoaling fish move towards the surface when attacked by predatory fish and marine mammals. Some approach the surface in shoals while breeding and this is particularly common in sandeels and Capelin. Perhaps most importantly, microscopic marine phytoplankton require sunlight for photosynthesis and so their abundance is very much higher near the surface. As a result, zooplankton follow them towards the surface in order to feed on them and fish that feed on zooplankton often visit surface waters. A complicating factor is that plankton often show a marked daily vertical migration in the sea, coming closer to the surface at night, with the result that predatory zooplankton and fish also approach the surface more at night than during the day. Yet the Black-legged Kittiwake probably does not feed extensively at night, although some individuals feed actively under the illumination from fishing vessels. However, they are active in late evening and at first light, and so take advantage of part of the nocturnal vertical migration of marine organisms. The Red-legged Kittiwake has larger eyes and feeds more at night, exploiting squid, lanternfish and other marine organisms, particularly those which glow (bioluminescence) at night.

FEEDING METHODS

Kittiwakes have been recorded using at least ten methods to obtain food.

1. Plunge-diving

This is the kittiwake's main method of feeding and involves plunge-diving from the air, usually while flying up-wind. The heights from which dives are made range from one to five metres, and the kittiwakes sometimes submerge for up to two seconds, but at other times they are scarcely submerged. This suggests that most of the food is obtained from the top half-metre of the sea surface and I suspect that the statement by Belopolskii (1957), and repeated by Baird (1994), that they catch food to a depth of one metre may be an exaggeration, but precise data do not exist. Typically, the dives are not vertical, as they are in terns, but also have a forward component, such as that often seen in gannets, and this arises from the flight speed followed by an instantaneous decision to dive without first reducing the forward speed. Kittiwakes do not hover over sites of potential prey, as is common in terns, but on occasions, they will quickly circle to return to a position where they had seen potential prey. This method is also described by Burtt (1974) who gives a detailed account of kittiwakes diving for food.

2. 'Phalaroping'

When the sea surface is calm or where solid structures prevent wave action, such as alongside ice faces or breakwaters, kittiwakes engage in what I have called 'phalaroping'. The bird settles on the water, and pecks at and consumes small organisms taken from the water surface, often spinning rapidly to select the next item in a manner reminiscent of feeding phalaropes. The objects are immediately swallowed and are not visible in the beak even with good optical aids, but I suspect they are small crustaceans, molluscs, small fish larvae and even insects which have fallen onto the sea.

3. Dip feeding

This method is usually used under relatively calm conditions. The bird patrols an area on the wing, at about 2–3m above the sea surface, detects a food item, swings round in a circle and picks the food item from the surface without landing on the water. Some nesting material is also collected by this method.

4. Surface dipping

This involves birds settling on the water and seizing prey below the surface by immersing the whole head. I have only seen this when a shoal of small fish is near the surface and where there are numbers of other seabird species feeding in the same immediate area. In my experience, this method is used much less than obtaining food while on the wing.

5. Diving from the sea surface

I have not witnessed this method of feeding, but it is recorded in the literature (Cramp & Simmons 1983), although this may be a misinterpretation of reports by others. It apparently occurs when seabirds aggregate over a surfacing shoal of fish, although taking food from the surface or just submerging the head without diving (surface dipping) is used much more frequently.

6. Feeding on the shore

Kittiwakes have been recorded feeding on barnacles on the intertidal zone in Newfoundland (Threlfall 1968). This seems to be an exceptional and surprising method and has not been reported elsewhere. The report of kittiwakes feeding on crabs in Baird (1994) might be a confusion with the use of 'crabs' as an all-embracing name for crustaceans in general.

7. Taking food from people

This is uncommon, but at Amble Harbour (Northumberland) and at one freshwater lake near the mouth of the River Tyne in north-east England, kittiwakes take bread thrown by the public for *Larus* gulls and ducks. They take the food while it is in the air or as soon as it hits the water surface by dipping while in flight in much the same manner as Black-headed Gulls. Kittiwakes have also been noted scavenging around harbours, feeding on dead fish and offal thrown into the water, but this is not a major source of food.

8. Commercial fishing offal

In many areas, kittiwakes aggregate and closely follow moving fishing vessels, taking offal thrown overboard while fish-gutting is taking place (Furness *et al.* 1992, Garthe & Hüppop 1994). Large gulls are frequently in attendance, but the kittiwakes follow closer to the stern of boats, and rely on their agility to obtain food while avoiding competitive scrambles with other species. Kittiwakes also attend fishing vessels while the nets are being hauled, when gutting is taking place and when under-size fish are discarded. However, in some areas, they are infrequent during such hauls, presumably because they are out-competed by larger gull species, Gannets and occasionally Fulmars and Great Skuas. When several species are present, kittiwakes that do manage to obtain food items are frequently pursued and have the items stolen. It has been suggested that kittiwakes in the North Sea feed on fishing offal more in winter than in summer, and offal appears rarely in the food given to their young. Many of the kittiwakes present in the North Sea during winter come from more northern breeding areas, where the habit of obtaining offal in this manner during the summer may be more frequent. Black-legged Kittiwakes have been recorded attending at flood-lit fishing vessels hauling nets at night, when competition from other species is less, but extensive feeding in the dark is probably infrequent except at the extreme northern end of the winter distribution, which in places can be beyond the Arctic Circle.

9. Sewage

O'Connor (1974) recorded kittiwakes feeding in numbers where sewage was released periodically into the sea near Howth Head in Ireland. I doubt if the kittiwakes were consuming sewage, but I suspect they were taking fish which were feeding on the sewage or, alternatively, brought to the surface by upwelling resulting from the mixing of fresh and salt water. When the tidal stretch of the River Tyne in north-east England was receiving untreated sewage (from over a million people) until the 1960s, kittiwakes were commonly present along the river, but they rarely fed there and they avoided the sewage outfalls frequented by other gulls (Fitzgerald & Coulson 1973). A few adult kittiwakes ventured further up the river into less polluted fresh water and captured small salmonids (Coulson & Macdonald 1962). Later, when sewage from the Tyne area was piped to a collecting area, processed and then shipped out to sea and dumped over 6km from the shore, kittiwakes fed abundantly over those areas where sewage was dumped, but again they appeared to feed mainly on small fish brought to the surface by the turbulence produced by mixing of fresh and salt water, rather than consuming sewage. In more recent years, the River Tyne has become much cleaner, but kittiwakes breeding on buildings alongside the river at Newcastle and Gateshead regularly fly down the river for 17km to reach the open sea, and only then start to feed. Almost all of the food fed to the young of these inland breeding kittiwakes was sandeels which had been caught in the open sea.

10. Dubious records of food

Apart from the above methods of feeding and the doubt already expressed about kittiwakes consuming sewage, Del Hoyo *et al.* (1996) list kittiwakes pirating food from alcids, terns and seals, taking birds' eggs, and consuming earthworms, small mammals, potatoes and grain. Belopolskii (1957) also records berries. If such methods occur, then they are very exceptional and are not normal sources of food consumed by kittiwakes.

FEEDING AREAS

Typically, kittiwakes hunt for food singly or in small flocks, although sometimes they collect in large numbers where abundant food has become available, such as where upwelling and tide-rips occur or where large shoals of fish are forced up to the surface by underwater predators. In most areas, kittiwakes feed offshore and out of sight of land. The distances moved by breeding adults at colonies are limited by individuals having to commute between the feeding grounds and the colonies, but this can still involve some individuals regularly travelling up to 60km and some to 90km or more from the colony to feed.

In a few locations, such as in Prince William Sound in southern Alaska, kittiwakes feed regularly in inshore waters, and even within sight of the colony (Irons *et al.* 1987), but this is exceptional. It is not clear why kittiwakes elsewhere do not feed near the shore to a greater extent. Near-shore areas often have large fish populations and these are regularly exploited by numerous terns, but usually they are ignored by kittiwakes, despite similarities in their feeding methods. In north-east England, I have seen kittiwakes plunge diving for food within 300m of the shore on very few occasions over a 50-year period and it appears that most individuals are feeding 10–90km from their colonies.

Apart from the protection offered by breeding on small islands, such sites also give the kittiwakes a greater amount of sea area from which to obtain their food, as the birds potentially can go in any direction to search for fish. Colonies on mainland coasts have only about half of this sea area available within their limited flying distance from the colony. The importance of the size of sea area available for feeding kittiwakes has not been investigated, but other factors, such as areas of upwelling and banks with appreciable fish stocks, probably also play an important role. So what is the best site for a kittiwake colony and what determines its upper size limit? These questions are still waiting to be answered.

FOOD IN THE BREEDING SEASON

Methods of reporting and comparing diets

Comparing the importance of food species taken by kittiwakes from published records is difficult because several different methods have been used. Some researchers have recorded the proportion (frequency) of stomachs or regurgitated samples which contained a particular food type. Others have recorded the total number of specimens of each species in the food samples, while yet others have gone further and taken the size of the items into account and then expressed the food in terms of the proportion of the total weight of food attributable to each species. The difference in these values derived from the same set of samples can be huge. In one case, 70% of the stomachs contained small crustaceans and these formed 85% of the individual items consumed, but because they were small, they contributed only 10% of the weight of food ingested. Most studies gave only one of these measures and they are not easily converted to the others. Another method which may be used more in the future is to convert each prey to its calorific content and express food items as a proportion of the total calorific content consumed (Jodice *et al.* 2008). Even this method is not without problems, since parts of the prey are not digested but contribute to the total calorific value of the prey obtained from bomb calorimetry. To be totally accurate, the amount and calorific value of the undigested material for each food species is needed and this is not likely to be obtained without keeping kittiwakes in captivity.

When considering the calorific value of food, some have simply taken the values for a particular species from other studies, although these are often based on a single measurement. A further variable which needs to be taken into account has been recently recognised, which is the seasonal change in the calorific (energy) content within each prey species and which differs between the sexes and with size and state of breeding (e.g. female fish containing eggs and different age classes). In addition, Wanless *et al.* (2005) have reported that there can be appreciable year-to-year variation in the calorific content of both sandeels and sprats, even among individuals of the same size, sex and stage of growth, and this probably applies to other prey species.

Quantitative studies on the diet of seabirds have tended to be an imprecise science, and the best studies are those which are comparative between the sexes, species, locality or year and which are made by the same investigator, because these allow strict comparisons. However, such studies are relatively few, but some are reported later.

Food selection

Throughout most of their range, both species of kittiwakes feed almost entirely on fish during the breeding season, typically capturing individuals shorter than 100mm long and swallowing them whole. Occasionally, fish up to 160mm have been recorded from stomachs, but large fish are difficult to handle and swallow, and are therefore usually ignored or rejected. This size limitation means that only the younger age classes of some fish species are available as food for kittiwakes, and this applies particularly to Herring, since the immature fish often have a more restricted distribution than the adults.

With the wind frequently ruffling the sea-surface, a kittiwake flying 2–8m above the water at about 40km/h has little opportunity to identify the species of fish which are below the surface before diving to make a capture. Thus there are major practical limitations as to how selective kittiwakes can be in selecting fish, and size may be one of the few choices the birds make. Despite often taking a high proportion of a single species in their food, kittiwakes are not specialist feeders and they feed on the fish species that are within their preferred size range and which are the most frequent in the surface waters.

This may seem to be at variance with the predominance of sandeels in the food of most kittiwakes breeding within the North Sea, but these fish have been estimated to compose 15% by weight of all fish in the North Sea. When larger fish and those which remain in deeper waters are excluded, sandeels form the great majority of fish available to kittiwakes.

When several fish species are taken, they are probably consumed in proportion to their suitable size and availability. This assessment is supported by the fact that kittiwakes frequently took the bony Snake Pipefish *Entelurus aequoraeus* (a relative of seahorses) when the population of this fish temporarily exploded in parts of the North Sea in the early 2000s (Harris 2006, Harris *et al.* 2008). These fish have rigid, bony bodies, and prove to be difficult to swallow and digest by both adult and young kittiwakes. Despite having a relatively low energy content, they were frequently taken by kittiwakes (and terns) when they were abundant. They also were frequently rejected when fed to kittiwake chicks and were commonly left on the nest before eventually falling into the sea. Some have suggested that when kittiwakes take pipefish, it is an indication that they are suffering from food shortage, and so are forced to take 'junk food'. I am not convinced that the capture of pipefish necessarily indicates a general food shortage, and consider it more likely that kittiwakes are capturing these fish simply because they had become numerous and they fell into the general size category of their prey. It remains to be shown that kittiwakes are selective in the fish species they capture, rather than just taking the available fish within a preferred size range.

In many areas, taking whatever fish are available still results in only one or two fish species dominating the food that is consumed by kittiwakes from a particular colony, but over a larger geographical distribution, many fish species are involved. In general, sandeels and Capelin are the most commonly taken fish, while Cod, Herring and Pollock are also frequently consumed in some areas. In most areas, small quantities of invertebrates are eaten, although these only become appreciable in the high Arctic. Belopolskii (1957) recorded small numbers of berries in the food of kittiwakes in the Eastern Murmansk region of Russia, and he and Braune (1987) also recorded insects, but neither state how these were obtained. Flying insects often drift long distances over the sea and sometimes fall and are trapped on the surface of the sea, while berries float and are frequently washed into streams and then into the sea, and I suspect both were picked up at sea, rather than that kittiwakes had been feeding on land.

The preference for fish in the diet probably reflects their availability and size. Fish tend to have a high calorific (energy) value due to the presence of oils, and since about 72% of the weight of a fish that is consumed is digested (Pearson 1968, Brekke & Gabrielsen 1994) (compare this with about 65% of the mass in crustaceans), nutritionally they are a good source of energy, but whether kittiwakes can respond to these differences and so be selective has yet to be studied.

Geographical variation in food

Geographically, there is a distinct south to north change in the species taken most commonly by kittiwakes. At the southern end of the kittiwakes' range, sandeels (called sandlance in North America) are the most frequent food fed to the young. The main species in the eastern Atlantic is the Lesser Sandeel *Ammodytes marinus* followed by the closely related *A. tobianus*, while in the western Atlantic, *A. americanus* is abundant. In colder waters off northern Norway, the Barents Sea and in Newfoundland, Capelin *Mallotus villosus* replace sandeels as the main food of kittiwakes. In the high Arctic, such as Svalbard and Novaya Zemlya at the northern limits of the kittiwake's range, Arctic Cod *Boreogadus saida* dominates as the main fish food and, in this area alone, small crustaceans, particularly species of *Thysanoessa* and *Mysis*, occur in huge densities where upwelling occurs at ice faces in the sea and become a major food source during the breeding season.

Records of the food of kittiwakes during the breeding season and in the Pacific Ocean are extensive, but few studies were spread over the whole breeding season and these also suffered from the use of different methods to evaluate their composition. Together, the studies showed a wide spectrum of food, which varied from area to area.

On both sides of the north Pacific Ocean, the sandeel *Ammodytes hexapterus* was the main food taken by kittiwakes in the southern part of their range. Only in Prince William Sound in southern Alaska were Herring *Clupea harengus* recorded as a major food and these and Capelin formed the main food consumed by kittiwakes (Baird & Gould 1986). Further north on Kodiak Island, sandeels and Capelin were the main food (Baird & Gould 1986, Baird 1990). Based on several studies in the Aleutians and the Pribilof Islands, Pollock *Theragra chalcogramma*, sandeels, Capelin and crustaceans featured as the dominant food species recorded in different years and on different islands (Springer *et al.* 1986, Schneider & Hunt 1987, Baird 1994). Sandeels and 'cod' were dominant food species in the Chukchi Sea area (Springer *et al.* 1984). In the high Arctic northern North Atlantic, both Arctic Cod and crustaceans appear to be the main food species consumed.

There can also be appreciable variations in the food taken by kittiwakes over quite short distances. For example in Scotland, sandeels are the dominant food of kittiwakes in the breeding season, but in restricted areas where sandeels are uncommon, such as the inner part of the Firth of Forth, other species of fish replace them as the major food source (Bull *et al.* 2004). In northern Norway, in years when Capelin were scarce, Herring replaced them as the main food (Barrett 2007b). It would seem that young Herring occur less frequently close to the surface than do sandeels and Capelin and so are taken less frequently by kittiwakes than would be expected from the size of their stocks. As a result, in years of Capelin shortage, kittiwakes switch to Herring but their breeding success is lower presumably because Herring are less easily obtained.

Seasonal changes in food

There is clear evidence that kittiwakes change the main species they take as food during the breeding season, as the abundance and availability of individual fish species within range of the colonies varies. For example in the Barents Sea, Belopolskii (1957) recorded changes in the dominant prey taken by kittiwakes from Capelin in the spring, changing to Herring and then to sandeels. Similarly, Lewis *et al.* (2001) found that on the Isle of May, Scotland, planktonic crustaceans were important early in the season, quickly followed by Sprats *Sprattus sprattus*, but sandeels dominated for most of the breeding season. However, the sandeels taken switched from the larger and older individuals in May to the smaller 0-group individuals (which had developed from eggs laid in the early spring) later in the season. In eastern and north-east England, Sprats come into inshore water in the winter and early spring and in this area they are the most important food for kittiwakes early in the year during the courtship period in March and April, but before egg laying. Later, there is a switch to sandeels, which are the dominant food given to the young until they become almost fully grown, when gadoids (cod family) tend to take over as the main food. In northern Norway and in Newfoundland, Capelin are the main food at the start of the breeding season and Herring often make an appreciable contribution to the diet later in the season, while in contrast, sandeels form only a very minor part of the diet throughout the breeding season. Arctic (or Polar) Cod and small crustaceans dominate the diet in the high Arctic, but it is not known how the use of this food varies during the breeding season.

Diet and feeding areas of the sexes

In some bird species, differences exist in the diets of males and females, but there is only one study on kittiwakes which gives enough information to allow a direct comparison of the food consumed by the sexes. This study was made by Belopolskii (1957) who examined the stomachs of 168 males and 130 females shot in the breeding season at Seven Islands, on the East Murmansk coast of Russia. He also sampled 21 males and 39 females shot further north in Novaya Zemlya. The results are shown in Table 2.1. While the diet in the two study areas differed, in each area the males and females consumed similar food and the small differences evident between the sexes can be attributed to chance variation within the samples. When the data collected by Belopolskii are analysed by the species of fish consumed (Table 2.2), the close similarity between the food of each sex still remains very clear, although again, the fish species taken changed markedly between east Murmansk and the high arctic Novaya Zemlya.

However, in the Lancaster Sound area, at 75°N, between Baffin Island and Devon Island in north-west Canada, Keith Hobson (1993), using the indirect stable isotope method of determining diet, suggested that in the breeding

TABLE 2.1 *The percentage of food items found in stomachs of male and female kittiwakes shot during the breeding season according to the data given by Belopolskii (1957). Note that the percentages are those based on the total number of food items recovered and not on the number of birds examined.*

Food	East Murmansk		Novaya Zemlya	
	Males	*Females*	*Males*	*Females*
Number of birds examined	168	130	21	39
Fish	71.7%	69.5%	50.0%	66.7%
Molluscs	13.2%	15.3%	15.4%	7.4%
Crustaceans	8.8%	11.4%	34.6%	25.9%
Echinoderms	0.6%	0%	0%	0%
Insects	3.2%	1.5%	0%	0%
Berries	2.5%	2.3%	0%	0%

season amphipod crustaceans supplied about 65% of the protein taken and much of the remainder was fish. In an earlier study made in the same general areas, Bradstreet (1976) found 93% to 98% of the kittiwake diet was Arctic Cod, so there may be marked differences in food collected there by kittiwakes between years.

Despite the similarity in the diets of male and female kittiwakes, there is circumstantial evidence that the sexes might show differences in the areas from which they capture food, with males tending to move further from the colonies to feed. Belopolskii (1957) first mentioned this possibility in a brief comment about the summer movement of Herring towards the coast from offshore areas and that they were first taken as food by male kittiwakes and that they only became part of the food of females later, when the fish had moved closer to the coast. Hobson (1993) also detected a sex difference in the food consumed by kittiwakes, reporting that 60% of protein consumed by males came from amphipods but that this reached 80% in females. Additional circumstantial evidence comes from the study at North Shields where the lengths of overnight feeding trips by males were longer than those by females, again raising the possibility that the males may have been travelling further from the colony. Another hint of differences in feeding areas comes from the high mortality in 1998 of breeding kittiwakes some 5 to 7km off the mouth of the River Tyne in north-east England, which was caused by a neurotoxin. Large samples of dead individuals were sexed and these showed a bias in the sex-ratio to 70% females. This sex bias in the mortality was confirmed in the next breeding season, when there was an appreciable shortage of adult females in the colonies and, although many males occupied nesting sites, they did not obtain mates and failed to breed. In this case, either the females were more sensitive to the toxin or a higher proportion of females were feeding closer to the shore, so bringing them into contact with the toxin. In the next few years, the use of tracking devices and geolocators, which are now just coming into use on birds the size of kittiwakes, should confirm or correct this impression.

TABLE 2.2   *The percentages of all fish species in the diet of male and female kittiwakes, based on the same samples as in Table 2.1, taken from the study by Belopolskii (1957).*

Food	East Murmansk		Novaya Zemlya	
	Males	*Females*	*Males*	*Females*
Sandeels	18.2%	18.1%	0%	11.1%
Capelin	23.4%	22.4%	0%	7.4%
Herring	32.4%	27.6%	0%	0%
Cod	23.4%	30.2%	23.1%	22.2%
Arctic Cod	0%	0%	76.9%	59.3%
Sticklebacks	2.6%	1.7%	0%	0%

Between-year variations in food

The food of breeding kittiwakes at individual colonies has been shown on occasions to vary markedly from year to year. Belopolskii (1957) first drew attention to this in the Barents Sea, where the proportion of fish in the kitti-wake diet varied between 70% and 95% in four breeding seasons, while molluscs complemented these differences and varied from 2% to 15%. The lowest proportion of fish was taken in 1947, and a similar decrease in the consumption of fish was found in other species of seabirds, both diving and surface feeders breeding in the area, indicating that fish abundance was much lower in that year. Similar between-year changes have been reported by Pearson (1968) who found that on the Farne Islands, sandeels increased in the food of kittiwakes from 46% by weight in 1961 to 69% in 1962, while Barrett (2007b) showed the amounts of Capelin in the food of kittiwakes in northern Norway varied from 1.1% in 1994 to 93.5% in 2002 and the differ-ence was mainly made up by the consumption of Herring (Figure 2.1). In contrast, Lewis *et al.* (2001) showed that the frequency of occurrence of sand-

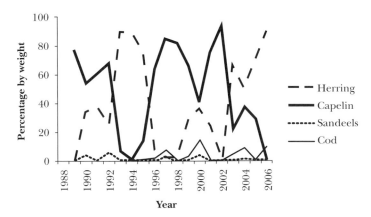

FIG. 2.1   *The percentage of different fish types in regurgitation samples from adults and kitti-wake chicks at Hornøya, North Norway between 1989 and 2006, based on data in Table 2 of Barrett (2007b). 'Cod' also includes other gadoids.*

eels in kittiwake food samples from the Isle of May from 1997 to 2000 fluctuated only between 87% and 93%, but there were greater differences in the proportions of 0-group and older sandeels consumed (Figure 2.2).

The food taken by kittiwakes has been found to vary markedly within a breeding season. This was first reported by Belopolskii (1957) who found that sandeels (12% to 31%)and Herring (6% to 48%) increased progressively in importance as the breeding season progressed, while capelin decreased from 38% to 1% between late April and September. Similarly, Pearson (1968) found that in 1961 sandeels increased from 42% of the food in June to 58% in July, but the reverse trend was evident in 1962. In both years, the proportion of clupeids (herring and sprats) increased towards the end of July, that is, near the end of the breeding season. More detailed studies of the consumption of sandeels (Lewis *et al.* 2001, Bull *et al.* 2004) have shown that the consumption by kittiwakes changed at the end of May in each season from one-year-old and older sandeels to the smaller 0-group individuals hatched that year from eggs and in their first year of life (Figure 2.2). They suggested that differences in the breeding success of the kittiwake between 1997 and 2000 could be attributed to the lower calorific density (energy content) of sandeels in their first year of life. However, this interpretation seems inadequate, since the difference in calorific values was small and, in all four years of study, breeding kittiwakes were feeding their young predominantly on the small 0-group sandeels in June and July when all of the chicks were growing.

In some areas, such as Shetland, kittiwakes appear to have no alternatives for adequate food sources, and rely entirely on sandeels as the main food source. Although Herring also occur in this area, they are only adult individuals which are too large for consumption by kittiwakes. When the sandeel population either moved away from the islands or crashed during the kittiwake's breeding season, the resulting shortage caused extensive breeding failures (Hamer *et al.* 1993, Heubeck *et al.* 1999).

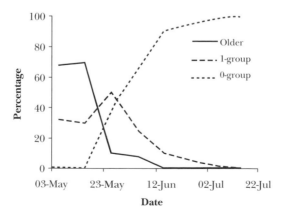

FIG. 2.2   *The proportion of three age classes of sandeels (Ammodytes) consumed by kittiwakes breeding on the Isle of May from 5 May to 14 July 1999 (after Lewis et al. 2001).*

Flying speed and feeding range

Flying speed contributes to the distance that kittiwakes can travel on feeding trips. Edward White and I measured flight speed of adult kittiwakes over a one-kilometre stretch of coast near a colony in north-east England. In relatively calm sea conditions and with light winds, individuals travelled at an average of 43km/h, but with moderate winds and a strong sea swell, kittiwakes tended not to fly in a steady direction, but swung into and out of the troughs, thus reducing their speed between two points to between 30 and 38km/h. Movements into strong, force 6 head-winds tended to reduce progress to below 25km/h. We did not measure flight speeds with moderate or strong tail-winds, but presumably these would exceed 43km/h. Pennycuik (1987) gives the flight speed of kittiwakes as 47km/h, but the figure reported by Götmark (1980) of 54km/h seems surprisingly high, unless it was wind assisted. Kotzerka *et al.* (2010) gives a wide range of flight speed based on GPS records, but the low values presumably include periods of inactivity. There was a peak at 40–45km/h, but their records of 85–90km/h seem unrealistic.

Colony attendance greatly limits the feeding range of kittiwakes, because daily or even more frequent returns to the colony are necessary, and most feeding ranges are based on indirect measurements based on lengths of absences from the colony. Pearson (1968) estimated an average maximum range of 55km, while Daunt *et al.* (2002) claimed a greater distance of 73km. In the Pacific, Kotzerka *et al.* (2010) tracked birds up to 59km from the colony. No doubt a few birds go further than this, while many travel shorter distances. There is the possibility that overnight trips also allow breeding birds to go further, but judging by their relatively late arrival back to the colony in the morning, I suspect these birds spend some time feeding at first light before starting the return journey. A major question remains to be answered: why do kittiwakes need to go so far to feed, when terns do not?

FEEDING IN THE WINTER

There are few small fish in the surface waters of the North Atlantic and Pacific Oceans during winter and kittiwakes probably switch to very different food during their pelagic winter distribution. There is very limited information on the food consumed by kittiwakes in the winter months, but clear evidence of a switch comes from the measurement of stable isotopes in kittiwake feathers to identify the trophic levels of food consumed. This method is valuable, but it does not identify the food species concerned. Analysis is made using primary wing feathers, as their composition reflects the type of food consumed while these were growing. Thus by comparing the stable isotopes of carbon and nitrogen in the outer primaries of kittiwakes which had grown in the late autumn and early winter, with those in the inner primaries taken from the same birds and which had grown in summer, Karnovsky *et al.* (2008) were able to show that the autumn and winter food of kittiwakes came from a lower trophic level than that consumed in summer. This suggests that in the winter, kittiwakes switch from fish to invertebrate food. There is little information on specific food of kittiwakes during the winter, but the few birds examined support this conclusion and indicate that in the winter, kittiwakes are feeding on invertebrates living in the surface waters of the oceans and particularly on highly specialised polychaete worms and molluscs that have flotation devices to keep them at the surface of oceanic waters. Among the several polychaete worms found in surface waters of the oceans, those of the genus *Tomopteris* have been found in the stomachs of kittiwakes in winter. They have large para-podia (paddle-like extensions of the body wall) which act as a means of locomotion and as buoyancy devices by greatly increasing their surface area to assist the worms to remain in the surface water. Each worm also has a light-producing organ which may attract kittiwakes to it in the half-light of dawn and dusk, or even during the night if and when kittiwakes feed in total darkness.

Another particularly important group that has been found in kittiwake stomachs during winter are the gastropod molluscs of the order Pteropoda (a word which literally means wing-foot) and which are commonly called sea-butterflies. They have large flotation 'wings' extending from the foot and these enable them to float and feed at the water/air interface. These molluscs are only 20–30mm long, but can occur in vast numbers in the oceans and sometimes form large aggregations. Some genera (such as *Clio*) have a calcar-eous shell. Concern has been expressed that slightly increased acidification of the surface water of the oceans due to increased carbon dioxide in the atmos-phere may cause the erosion of the shell and the earlier death of individuals, thus reducing their numbers. Not only would this have an impact on kitti-wakes, it could also lead to a major reduction in the amount of calcium carbonate (in the form of aragonite deposited in the shells of dead ptero-pods) falling to and then being deposited on the sea bed and thus removing even less carbon dioxide from the atmosphere (Blank & Gruber 2007). A reduction in pteropods could also lead to a decline in other surface-living marine invertebrates that may be utilised by kittiwakes. Unfortunately, studies on the numbers and abundance of small oceanic-living invertebrates, particu-larly in winter, are extremely few.

In the Pacific Ocean, the stomach contents of kittiwakes shot while wintering off the coast of California contained squid *Loligo* sp., Anchovy *Engraulis mordax* and small quantities of crustaceans, mainly *Thysanoessa*. The food of kittiwakes elsewhere in the Pacific Ocean during the winter does not seem to have been investigated, but it is likely that invertebrates also play an appreciable part in their diet, as they do in the Atlantic Ocean.

Little is known of the abundance and winter distribution in the oceans of potential food sources for kittiwakes, nor how much their numbers change from year to year. What is evident is that when there are strong winds, many of these animals sink from the sea surface into more stable waters several metres below the surface, thus becoming unavailable to kittiwakes, and only return there when the wind and sea conditions moderate. While kittiwakes may find oceanic conditions allow them to avoid predators, food availability probably fluctuates violently. It is essential that kittiwakes move away from atmospheric depressions and associated strong winds which cause rough seas, and move to areas where successful feeding is again possible. The edge of some of these movements which involve many thousands of kittiwakes have been observed from coastal headlands in Europe. The kittiwakes stream along the coasts in flocks of up to fifty birds passing a few hundred metres offshore every few minutes. They are usually flying at a right angle to the strong wind and rather than flying in a steady line, they dive and rise in and out of troughs caused by the swell. The continuation of these movements eventually takes the birds away from strong cyclonic winds around atmospheric depressions. Much more will be known about how kittiwakes exist in oceans when position loggers are small enough to be employed on them without being an unacceptable load on the individuals concerned.

How much food does a kittiwake need?

In recent years, Gabrielsen and his co-workers have made extensive studies of the energy requirements of kittiwakes. Their needs are calculated in terms of energy and the results are given in kilojoules (kJ), an adult in Svalbard requiring 769kJ per day (Gabrielsen *et al.* 1987). For most people, this figure will have more meaning if converted to the consumption of fish, although this cannot be very precise because the calorific value of fish varies with species, year, size and breeding condition. When not breeding, it has been estimated that each kittiwake requires approximately 315g of Capelin or sandeels per day and this represents about 80% of the body weight of a kittiwake. Even more food is required when rearing a chick, and this has been estimated at 419g of Capelin a day, which is about 105% of the body weight of an adult in the high Arctic where the study was made (Gabrielsen *et al.* 1992). Although not calculated, it is obvious that the food requirement is even higher when a brood of two chicks is being raised and presumably the requirement for each parent and half of the needs of the chicks could reach 120–130% of the body weight of each parent.

Food of the Red-legged Kittiwake

At some localities in the Pacific Ocean during the breeding season, both kittiwake species have been observed feeding together from the same shoal of fish and there is an overlap in the prey species taken, although this may be small. Transects at sea off the Pribilof Islands and the Aleutian island chain suggest that feeding trips in the breeding season by the Red-legged Kittiwake extend further from the colonies than for Black-legged Kittiwakes, and that individuals of each species from the same mixed colony may use appreciably different areas for feeding. Storer (1987) suggested that the larger eyes of the Red-legged Kittiwake are an adaptation for night feeding. This is supported by stomach analyses of the two species during the breeding season and also by the food regurgitated by chicks at the nest where the red-legged species consumed much more lampfish and squid, both of which are deep-ater species with a pronounced vertical migration from deeper water to the surface at night. In the Pribilof Islands, more than half of the stomachs of Red-legged Kittiwakes examined contained the Northern Lampfish *Stenobrachius leucopsarus* and these were much less common in the stomachs of Black-legged Kittiwakes (Hunt *et al.* 1981, Dragoo 1991, Lance & Roby 1998). Both lampfish and many species of squid have bioluminescent organs, and the light they produce at night may play an important part in the discovery of this prey by Red-legged Kittiwakes.

The most valuable comparison between the food of the Black-legged and Red-legged Kittiwake was made by Lance and Roby (1998, 2000) because they obtained samples of the food simultaneously for the two species in the same colony on St George Island of the Pribilof group of islands. They measured the proportions of different food items by volume from samples regurgitated by young in the nest. The difference in the diet of the two species is large and is illustrated in Figure 2.3. In addition to the differences shown in Figure 2.3, the Black-legged Kittiwake chicks received about 10% fish offal in their food, which was absent in the Red-legged Kittiwake food. In addition, the deep-water shrimp *Gnathophausia gigas* occurred in the food of the Red-legged Kittiwake but was not fed to Black-legged Kittiwake chicks. Overall, the food Red-legged Kittiwakes fed their chicks contained more lipids, but this only marginally increased the calorific value of the food.

It is possible that the dependence of the Red-legged Kittiwake on being

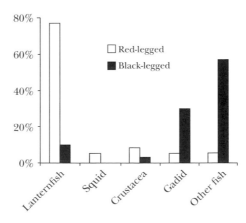

FIG. 2.3 *The composition of food fed to young by Black-legged and Red-legged Kittiwakes on St George Island, Alaska in 1993. Based on Lance & Roby (1998).*

able to reach areas of deep oceanic water from their colonies, so avoiding competition for food with the Black-legged Kittiwake, plays an important part in determining their very restricted distribution in the North Pacific. The relatively shallow and extensive continental shelf in areas where many Black-legged Kittiwakes breed would be unsuitable for Red-legged Kittiwakes because they appear to select deep water and these areas may occur too far from colonies of the Black-legged Kittiwake.

Nothing is known of the winter food and very little about the winter distribution of Red-legged Kittiwakes in the Pacific Ocean. The few records suggest that they winter well offshore, and probably remain over deep oceanic water.

FOOD AND BREEDING SUCCESS IN THE BLACK-LEGGED KITTIWAKE

There are at least two important factors relating to their food which can influence the breeding success of the kittiwake. Firstly, the abundance of fish utilised by kittiwakes can vary dramatically from year to year. Secondly, the number and abundance of secondary species of fish which are available as alternative food sources vary from area to area. For example, Barrett (2007b) showed that while Herring were abundant off the north Norwegian coast when the Capelin stock there crashed, herring were not taken in large enough numbers by kittiwakes to totally replace the virtual absence of Capelin in the diet and this resulted in the breeding success of kittiwakes being markedly reduced in these years. This was not because the stocks of Herring were inadequate (they were demonstrated to be vast), but rather because the kittiwakes had more difficulty exploiting them than Capelin, presumably because they approached the surface less frequently.

In some areas, such as Shetland, the shortage of sandeels in certain years resulted in poor breeding success in kittiwakes and terns. It would appear that in areas where sandeels were scarce, neither abundant nor adequate alternative fish species were available. In contrast, the breeding success of kittiwakes around the River Tyne area of north-east England has remained high every year from 1953 to 2010, indicating that food shortage during the breeding season in that area was extremely rare. In general, there has not been a widespread decline in food for the kittiwake, but during the 20th century the availability of food has varied sporadically and sometimes differently even between colonies within relatively short distances of each other.

It is reasonable to envisage a range of marine areas used by breeding kittiwakes. The marine habitat is variable, and there are areas with and without adequate alternative foods for kittiwakes. Where alternative food species are few or absent, periodic breeding failures of kittiwakes are likely to occur. Some areas have alternative and reliable sources of food for kittiwakes during the breeding season, and so breeding success is consistently high because even if one source fails, others are available and an overall food shortage does not occur. The area off the mouth of the River Tyne in England would seem to be an excellent example of this type of habitat. Other areas, such as Shetland, may not consistently offer alternative fish species to be exploited, and breeding success of kittiwakes might be expected to fluctuate markedly in years where their main food source becomes inadequate.

Are there areas which may frequently have unsuitable or insufficient fish stocks to support breeding kittiwakes in most or even all years? In these areas, of course, breeding kittiwakes are likely to be absent. Such areas may possibly exist along most of the Labrador coast, in east Greenland, and in sections of the Arctic coast of Russia. Are such food-deficient areas a factor (one of several) in determining the distribution of breeding kittiwakes? In the Pacific Ocean, perhaps areas in the warmer waters to the south of the current distribution of kittiwakes may also be inadequate.

These considerations explain why breeding failures in kittiwakes have not been reported in some areas, but they do not account for the more frequent failures in recent years. One possibility is that the trend of increasing breeding failure in kittiwakes is, at least in part, an artefact arising from the much more intensive observation of colonies in recent years and that cases of breeding failure in the far fewer colonies which existed 50 to 100 years ago simply went unnoticed and unrecorded.

There are now very many more people, compared with those fifty or more years ago, observing seabird colonies, and census work has increased dramatically. There are more wardens observing and protecting colonies, and more annual reports now record detailed data. Centralised and systematic recording of breeding productivity has been developed in many countries, but only in recent years. For example, we have little knowledge of the breeding success of kittiwakes in most British colonies in any of the ten years from 1950 to 1959. In those years, very few colonies had wardens or other observers whose knowledge of the colonies spanned many years and even fewer recorded information. On the Farne Islands, occasional poor breeding years for kittiwakes have only been reported in recent years. The National Trust of England has employed wardens there annually since the late 1940s. However, both the knowledge and duties of the wardens have changed over time, and annual monitoring of the seabird numbers and their breeding success has only been introduced relatively recently. In the earlier years, the wardens' main duties were to collect landing dues and to stop people entering tern and eider breeding areas, and they were not required to produce written reports. In at least two years, the wardens were monks – they were delightful people, but their natural history knowledge was minimal.

Another possibility is that the fish stocks have declined markedly in recent years, perhaps as a result of excessive overfishing by humans, or even as naturally occurring events. Stocks of some commercial fish have been monitored over many years and the size of stocks has been shown to have changed dramatically in recent years. Perhaps these are best demonstrated by the changes in stocks of Cod in the North Sea, which have not only declined over time, but the average size and age of individuals have decreased dramatically. In the early 1900s, there were many giant, old Cod in the North Sea, which were predators on smaller Cod and other fish. As fishing intensified, many of the large Cod were removed and they were replaced by many more but smaller individuals which tended to maintain the biomass of the stock. Exploitation continued and increased, the average size of individual cod continued to decline, and the age of first spawning by females became lower, while the whole population dynamics of the species changed. In the early 1900s, the increasing exploitation of the Cod stock probably increased the numbers of small fish, including Cod, Herring and sandeels, as the intensity of natural predation on them declined. Thus the early days of human exploitation of Cod had no adverse impact and the more recent exploitation until about 1920 may have even assisted the increase of many seabird species breeding around the British Isles during the 20th century, including the kittiwake. The increasing exploitation of fish in the North Sea, including Herring, Haddock and Cod, has probably changed the size and structure of both commercial and non-commercial fish populations, but the effect on such key species as sandeels and sprats has not been intensively monitored or even investigated until recently.

It has been suggested that the introduction of commercial sandeel fisheries around Shetland and later in the Firth of Forth in Scotland was directly and mainly the cause of decline in the breeding success of the kittiwake not only in Shetland and the Firth of Forth, but elsewhere. The evidence for this is, to me, still ambivalent. Breeding by kittiwakes (and other seabirds) has failed in many areas where commercial sandeel fisheries did not occur. In some places, the breeding success of kittiwakes was lower in some of the few years with an appreciable commercial sandeel industry in Shetland and the Forth, but the cause-and-effect relationship between the two has not been clearly established. Doubts exist because the extent of breeding failure in the kittiwake does not appear to correlate with areas where fishing was most intense, and failures occurred in some areas where no known fishery existed within the feeding range of breeding kittiwakes. Further, no good correlation exists between the intensity of fishing in different years and the adverse effect on kittiwake breeding success. The evidence mainly relies on simple correlations with year, and this relationship could have other causes. So far, the impact arising from sandeel fisheries on kittiwakes has not been satisfactorily separated from other, perhaps natural changes in the size and migration of sandeels in the areas where breeding failures occurred. However, the fishery was banned in restricted areas after a few years of exploitation because it was unregulated and harvesting sandeels for non-human consumption is undesirable on other grounds. It is unlikely that the true impact of this brief industry on kittiwakes will ever be firmly established.

The sizes of non-commercial fish stocks remain unknown over most of the range of kittiwakes. Determination of the status of sandeels and even Mackerel was not attempted around Britain until recently. There is similar lack of data for most of the North Atlantic and North Pacific Oceans and their coastlines. We simply do not know directly how stocks of many marine fish have changed during the past hundred years, and much of our knowledge depends upon records of commercial catches and the effort put into them. This lack of precise information makes it difficult to predict what might happen in the future.

Most fish have a small larval stage, hatching from eggs that are laid in very specific habitats (spawning grounds). The larvae which develop into young fish are weak swimmers and both stages live in surface waters which are drifted by currents and winds. In some years, this can result in young fish not reaching, or being less numerous in certain areas of the sea even without a population crash. Add to this the fact that breeding terns and gulls have a very limited feeding range from their colonies, with most terns feeding within 15km and kittiwakes within a radius of 60km of their colonies. Taking both of these factors together, this can mean that in some years and in some areas, breeding birds were unable to reach and exploit a food source which they had used in previous years because the fish distribution has changed, sometime by only a few tens of kilometres. This leads to the conclusion that the exploitation of a food resource where the predator can reach the prey for only a limited time and is only a limited area is a high-risk strategy. It is therefore not surprising that, from time to time, seabirds such as kittiwakes and terns encounter food shortages in the breeding season and breeding failures occur. While some of these failures may be induced by commercial over-fishing, it should not be unexpected to find that failures can and do occur naturally and without human intervention.

Natural shortages of food for kittiwakes and terns are likely to be local effects, and appreciable variations in breeding success can be expected between colonies and regions. Occasional years of breeding failure in terns and gulls are not of major importance to the population since adults are long-lived, and the deficit in young production can be made up in other years. Problems concerning the viability of the colony and species arise when breeding failures occur frequently and over a large area, as the deficit cannot be made up by other years of successful breeding or by immigration of young breeders. In some species, e.g. Herring Gull, alternative food sources such as landfill, agricultural land and intertidal areas offer means of moderating a shortage of fish food. The kittiwake does not have such alternative feeding methods.

CHAPTER 3

Winter oceanic distribution and reoccupation of the colony

OCEANIC DISTRIBUTION

KITTIWAKES spend much of the winter half of the year in the northern oceans, far from land and most humans. Observations made from ships during transatlantic crossings showed that in winter, kittiwakes were spread across the whole of the North Atlantic, the North Sea and the western end of the Mediterranean Sea, occurring regularly as far south as 40°N and sometimes reaching 30°N (Wynne-Edwards 1935, Rankin & Duffey 1948). There are also sightings as far south as the Tropic of Cancer, but records south of the equator are few and there are no ringing recoveries south of the equator. Nevertheless, Maclean (1993) suggested that some kittiwakes are regular visitors off Cape Town in South Africa between December and February.

In the Pacific Ocean, there are no ringing recoveries from the deep oceanic wintering areas, although some individuals are frequently seen as far south as 30°N and particularly off the coasts of Mexico and California, but these are just the edge of the extensive oceanic distribution. I have not found any systematically collected winter records of kittiwakes at sea within the Pacific Ocean.

The northern edge of the winter distribution of kittiwakes is less well known. In winter, kittiwakes are common at sea off the whole of the Norwegian coast as far as the North Cape, but they appear to be uncommon within the Barents Sea. On the western side of the Atlantic Ocean, northern limits of wintering birds are restricted by the distribution of ice between Canada and Greenland. Some kittiwakes occur up to the limits of open water in both the Atlantic and Pacific Oceans, but winter densities at this northern edge are low. That kittiwakes occur so far north in winter is surprising, since some are north of the Arctic Circle and so are in areas of total darkness, while others, a little further south, encounter very short periods of daylight.

Ringing recoveries are of relatively little use in evaluating the winter distribution of kittiwakes, since there are no recoveries in deep oceanic areas, although observations from ships show they are numerous across the whole width of the North Atlantic (and probably the North Pacific). The reason for this is that there are no humans capturing or finding dead kittiwakes in these vast areas of the oceans.

The recoveries which exist are mainly the result of hunting kittiwakes near the coasts of Greenland and Newfoundland (and the nearby Grand Banks) and to a lesser extent in the Bay of Biscay areas of France and Spain. However, these recoveries offer useful information when divided to give the monthly distribution of recoveries on both sides of the Atlantic Ocean north or south of 50°N. Immature kittiwakes from Britain move progressively south (and west) within the Atlantic during the autumn and winter months (Figure 3.1),

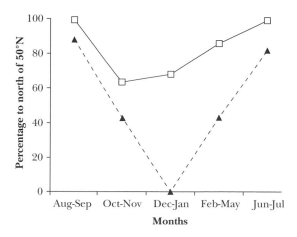

FIG. 3.1   *The percentage distribution of recoveries of immature kittiwakes north of 50°N in the North Atlantic, divided into monthly periods. The open squares and continuous line are for the European coast and the triangles and dashed line are for the western coast, mainly Canada and Greenland. Based on Coulson (2002).*

but with a greater proportion south of 50°N on the western than on the eastern side of the Atlantic. This difference is probably real, influenced by low sea temperatures induced by the cold Labrador Current carrying icebergs along the coast of Canada as far as Newfoundland, in contrast to the less cold sea temperatures in the eastern part of the North Atlantic as a result of the Gulf Stream.

The main spread of kittiwakes into the pelagic zone begins in early October, and birds reach their southernmost distribution in December and January, after which time they appear to slowly return north (Figure 3.1). The only other seabird encountered in numbers in this pelagic zone is the Fulmar, but this species has its southern winter limit farther north than that of the kittiwake.

In winter (November–January inclusive), there were only three adult kittiwakes out of 58 recoveries (5.2%) on the western side of the Atlantic, while the proportion was ten out of 105 (9.5%) for immature birds. This may indicate that adult kittiwakes do not travel as far in winter, but since some make transatlantic movements, there is an alternative explanation, namely that the adults have a similar oceanic distribution but spend less time each year in the pelagic zone (and so produce fewer transatlantic recoveries) because they return earlier to the inshore waters and their colonies.

Kittiwakes ringed in northern Russia have proportionately more recoveries reported in Greenland than in Canada that the British recoveries, and this suggests that there is considerable mixing of kittiwakes from other areas in winter, but on average, the birds from Arctic areas tend to occur further north than the British birds (Coulson 1966a). There is an increased range of wing lengths reported for dead kittiwakes washed ashore on North Sea and French coasts during the winter, which frequently include wing lengths greater than those of birds that nest locally. Many of these large kittiwakes had not completed their primary moult by January, whereas the smaller individuals had fully grown primaries by this time, so the stage of moult may also be a useful clue as to their origins. Much more information could be obtained from the systematic examination of the size and moult condition of winter casualties.

MOVEMENTS OF KITTIWAKES

Do kittiwakes migrate? The answer to this question depends much upon how 'migration' is defined. The centre of kittiwake distribution certainly changes between summer and winter and all individuals move some distance from the colony in which they breed. All kittiwakes from the Barents Sea move in the same western and then southern direction into the Atlantic Ocean, but these directions are forced on them because they have to leave the Arctic Ocean and this can only be achieved by moving in an overall southwesterly direction. Kittiwakes from the west coasts of Europe move west and south, while those from Newfoundland move south and probably east. In all of these cases, the direction is determined by the distribution of suitable ocean areas. Of course, such limitations on the directions of movement also apply to many land birds and in most cases the term migration is readily applied.

There is no evidence that individual kittiwakes winter near the colony in which they breed, nor that they move to the same area in successive winters (unlike Black-headed Gulls), although information on these points is very limited. It is also evident that irregular large-scale movements of kittiwakes are made to avoid adverse weather conditions in winter, causing individuals to range irregularly over large areas, and the term 'nomadic' would be appropriate.

In marked contrast to the winter distribution, the extensive mid-Atlantic pelagic zone is totally vacated by kittiwakes of all ages by mid-May, when numbers are concentrated over the shallow continental shelves around islands and coasts. This change to a coastal distribution is associated with a major change in the food of the birds and the greater consumption of fish.

One-year-old birds, with their characteristic black and white 'tarrock' plumage, also come into the coastal areas but do not visit colonies and often aggregate during the day with older birds on beaches, piers and rocky outcrops along the coast, while they complete their first annual moult, which takes place earlier than in the adults. Each night, these and older non-breeding kittiwakes move many kilometres out to sea, presumably to sleep and feed before returning to the shore in late morning. It might be thought that the movement of these immature birds into coastal waters indicated a return towards the areas where they had hatched in previous years, but this is often not so. Many of these young birds are thousands of kilometres from 'home' and, for example, some reared in Europe remain around the coasts of Newfoundland and Greenland. There is little evidence to demonstrate where young kittiwakes reared on the Atlantic coast of North America spend the summer, as relatively few have been ringed and there is even less indication of where immature kittiwakes move within the expanses of the North Pacific Ocean.

Many second-year kittiwakes remain at sea for most of the year. Some make brief visits to colonies, mainly in June and July, but even then they only visit during fine weather, and leave the colonies each night and fly offshore. An appreciable proportion remain far away from their natal colony, and again there are ringing recoveries of European birds spending part, if not all, of the summer on the opposite side of the Atlantic.

Some three-year-old kittiwakes breed for the first time, but in Britain these do not arrive at the colonies until April or early May, several weeks after the older adults. Fewer of the three-year-olds remain far away, but there are still a few recoveries in May and June of these birds over 1,000km from their natal areas. Not all return to breed in or close to their natal colony.

Prior to 1965, it was exceptional in Europe for kittiwakes to be seen on or close to the shore from October to January, and most records were of small flocks following the coastline, a kilometre or so offshore during extensive weather movements. Since then, small numbers of kittiwakes have occurred during winter in various coastal areas of western Europe, feeding and resting around harbours and fishing ports. These are probably not local breeding birds and, for example, I have not encountered any of my colour-ringed birds among them in north-east England, although I have seen some ringed birds from French colonies and other individuals seemed large and possibly from more northern colonies.

BEHAVIOUR, BIOLOGY AND MORTALITY AT SEA

Very little is known about the biology, behaviour and risks of mortality of kittiwakes during periods of pelagic existence. Over 80% of my marked adult kittiwakes disappeared and presumably died between the last time they were seen in the colony in August and when they would have been expected to return to the colony in the following spring, so it would seem that this pelagic life is not without problems. Winter mortality while far from land would explain why the ringing recoveries of kittiwakes are effectively about 1% of those ringed (after excluding records of those found dead soon after ringing as chicks).

So why do many kittiwakes die while far out in the oceans? There are very few predators of kittiwakes in offshore areas, and being able to avoid most predators is almost certainly the key advantage obtained by leading a pelagic life. The main causes of death when far from land and predators would appear to be starvation and disease. Pelagic kittiwakes do encounter a series of problems. They need food, they need to be able to rest and sleep at sea rather than on land, they have to manage without fresh water to drink and they have to cope with severe storms and periods of storm-force winds.

The lack of access to fresh water in oceanic seabirds is solved in all species by the well-developed nasal glands that are situated in skull depressions just above the eyes. These glands are able to excrete a salt solution which is about twice the concentration of that of sea water (Schmidt-Nielsen 1960, 1979). So by drinking two units of sea water and excreting one unit of this strongly saline liquid, seabirds are able to retain one unit of fresh water within their bodies. The excretion of this strongly saline solution can often be seen dripping from the bill-tip of kittiwakes and is not, as many people think, the result of recently dipping the beak into sea water. This process of salt excretion has energy costs to birds and it is not surprising that during the breeding season many, including the kittiwake, choose to drink fresh water from lakes near the coast or where small streams flow into the sea, but these are not always available.

How do kittiwakes find adequate nourishment in the pelagic zones, when daylight is reduced to about a third or even less of each 24 hours during the winter? In winter, some kittiwakes remain as far north as the Arctic Circle, so how do these birds feed and survive in the total darkness? The answer is not known. Some may feed behind fishing boats using floodlights, while others may be attracted to luminous prey.

What is already clear is that feeding must be impossible during severe winter storms, when the sea surface is whipped into a froth and waves and swell several metres high are produced. During severe gales, kittiwakes settle on the sea (Rankin & Duffey 1948), but I doubt if they can feed at this time. To be trapped in such storms must be a threat to survival, and I suspect that they often make efforts to avoid being caught in severe conditions by making long-distance movements away from the track of depressions and strong winds. Evidence that such movements occur is frequently witnessed on the coast of Britain, where thousands of kittiwakes (and other seabirds) have been seen to move hundreds of kilometres along the coast and in the same direction, leading them away from the depression. Rankin & Duffey (1948) also recorded movements of large numbers of kittiwakes flying in the same direction when in mid-ocean and, in some cases, over 100m above the sea surface. Most of these movements involve the kittiwakes flying at a right angle to the direction of the wind (i.e. north in an easterly wind), with the birds swinging in and out of long troughs between the swell. Since winds move around depressions in an anticlockwise direction, such flights will take the birds away from the strongest winds and eventually into areas where the sea conditions are less severe. Within 24 hours, a kittiwake could move about 1,000km and out of the typical west to east path of the depression. This would usually be far enough to avoid the worst of the strong winds associated with depressions and allow the birds to reach areas where they could feed. It seems likely that such movements form a regular part of the winter behaviour of the kittiwake and such movements make their oceanic distribution fluid and variable.

Prolonged storms in winter have caused some kittiwakes to be driven far inland (or 'wrecked'). About half of these are first-year birds and the remainder older birds, including adults, and many, but not all, are subsequently found dead. It is likely that many kittiwakes reported inland are individuals which have exhausted their food reserves, while others may have been ill and in poor condition before encountering the strong winds and were unable to continue long-distance movements. When inland, they are often reported on lakes and reservoirs, where they seem to find it difficult to obtain food and many are probably unable to recover their body condition and die.

Storms are not the only factor causing kittiwakes to occur inland, and there are a few fascinating examples of groups deliberately flying high and crossing mainland Britain in good weather conditions, usually moving from the North Sea towards the Atlantic.

In winter, many kittiwakes move in small flocks. The composition of flocks and for how long membership persists is poorly understood. Some flocks include both adults and first-year birds, but I doubt if there is kinship between the individuals, as pairs and their offspring separate once they leave the colony in the autumn. The parent–young bond is severed when kittiwake chicks fledge and finally leave the colony, and there is no post-fledging care of the young away from the colony (unlike that found in several other gull and tern species). I have often found considerable differences between the dates when each member of an adult pair was last seen in a colony at the end of the breeding season and also in the dates on which they return in the spring, and this adds further circumstantial support to the suggestion that the pairs do not stay together throughout the winter. While some pairs persist over several breeding seasons, I doubt if they retain contact with each other in the winter, wandering the oceans as individuals, forming temporary flocks and mixing with a multitude of other kittiwakes from many areas.

USE OF GEOLOCATORS

Geolocators are ingenious, small devices (1.4g) developed by the British Antarctic Survey. Because they have now been miniaturised, they can be used on kittiwakes without handicapping the birds. They record the time and length of daylight each day, so the actual time relative to the Greenwich Mean Time of dawn and dusk is recorded wherever the individual kittiwake happens to be. In winter a sudden change to shorter daylight indicates that the bird has moved further north, while a shift in the actual time of sunrise and sunset measures the extent of latitudinal movements with, for example, these both being later if the bird has moved west. Each geolocator is attached by a harness to the back of the kittiwake or to a leg ring when the birds are captured in the breeding season. Hopefully, they are retrieved in the following breeding season and the accumulated information is downloaded and analysed to record the geographic positions and movements throughout the year (Kotzerka *et al.* 2010). The technique is not as accurate as that obtained by the use of transmitters which send signals via satellites, but geolocators are cheaper and smaller and because they are light, they are more appropriate for use on kittiwakes.

In the last few years, geolocators have been attached to kittiwakes in several colonies in Britain and no doubt elsewhere and a huge amount of information is being obtained, but interpreting the information is time-consuming and the first of what will presumably be a series of publications has only just appeared (Bogdanova *et al.* 2011), in this case from kittiwakes marked on the Isle of May in Scotland. The data obtained confirm and expand the information already obtained from ringing recoveries, i.e. that kittiwakes from the same colony disperse widely over the North Atlantic in winter, making large, apparently wandering movements, with some reaching Canadian waters but others staying closer to Britain on the eastern side of the Atlantic. The interpretation of the results made by Bogdanova *et al.* (2011) suggests that the extent and direction of these movements are correlated with whether breeding had been successful or failed during the previous summer, with some failed breeders travelling further and in a different direction than individuals which had bred successfully. This may well be the case, since it is already known that failed breeders behave differently from successful breeders and often leave the colony earlier and return in the spring at a more variable date (Coulson 1966b, 1972).

However, an appreciably higher proportion of young breeding kittiwakes fail to breed successfully (Chapter 6) and the possibility exists that the sample of failed breeders was dominated by young birds. As a result, the extent of the movements could be an age effect rather than one related directly to breeding success, since young adult kittiwakes spend more time each year in the pelagic zone and so have more time to make extensive movements. I recorded an extreme example of this during the 1974–75 winter, when most old kittiwakes were absent from the colony for only ten weeks, while those which had bred for the first time did not return until they had been pelagic for over six months.

The study by Bogdanova *et al.* (2011) suggested that the trans-Atlantic movements were made by males, but one ringed female kittiwake from North Shields was recovered in Nova Scotia, Canada, during the winter following her first breeding attempt and it is evident that, on occasions, both sexes make trans-Atlantic movements.

Clearly the use of geolocators is a great advance in understanding the movements of kittiwakes and these will become even more valuable if and when they can be coupled with a detailed knowledge of the age or previous breeding experience of individuals. They will soon give a much greater insight into the winter movements of kittiwakes in the pelagic areas of the North Atlantic. In addition, more extensive information on the feeding areas used by adults in winter and during the time they are restricted to colonies during the breeding season will become available, filling an area of kittiwake behaviour which currently is not well understood.

THE ANNUAL REOCCUPATION OF THE COLONY

Kittiwake colonies are deserted in the autumn, and most of the seabird cliffs remain silent and empty until the late winter or early spring. In more southern areas, the cliffs are visited by returning Northern Fulmars as early as November, and there are irregular and infrequent appearances of Common Guillemots on the nesting ledges. The earliest I have found kittiwakes back at a colony in England was on 17 December, but this was exceptionally early and the reoccupation dates varied considerably from year to year even in the same colony. Throughout Britain, most colonies are not visited before mid-February, and the return is much later in colonies in the Arctic and sub-Arctic, where the first nest sites are not occupied until April.

There is no published literature that systematically records the timing of the return of kittiwakes to their colonies anywhere within their worldwide distribution, and the account given below is the only detailed account of the annual reoccupation of colonies. This is a huge gap in the ecology of the kittiwake (and some other seabirds), because the return and pre-breeding activity influence the time of breeding, affect clutch size and are important factors in determining potential productivity, as will be shown later. No doubt this lack of information is due to the difficulty of gaining access to kittiwake colonies in the first few months of the year because of inclement winter weather and the markedly short time spent at the colony when the kittiwakes first return (see below). Fortunately, viewing the kittiwake colonies at North Shields and Marsden in north-east England was relatively easy due to good road access, and extensive records have been collected of the return to these colonies over many years.

The presence of ice and snow on cliff ledges can make access to the nesting sites impossible for kittiwakes, and this is often the main factor which limits the time of return in the Arctic. Everywhere, the wind speed is an important factor influencing the day-to-day presence and numbers of kittiwakes at a colony early in the season. The age and sex of individuals, as well as the geographical location of the colony, and even the size of the colony, all influence the time of return of kittiwakes. In addition, the time of reoccupation of all colonies in north-east England has changed dramatically by over two months in the past 50 years. These factors are considered in more detail below.

STAGES IN THE ANNUAL REOCCUPATION OF THE COLONY

The reoccupation of the colony by kittiwakes is a gradual process, and I have identified eight stages which were spread over several months, but they can be telescoped into a shorter period if the initial return is late. The dates given below for each stage were recorded at Marsden, Tyne and Wear (55°N), in several years in the 1950s and early 1960s, when intensive and frequent observations were made almost daily.

Stage 1 (3–15 January). The first evidence of imminent reoccupation of the colony is a small flock of five to twenty adult kittiwakes forming a compact raft on the sea some 100–200m or so away from the cliffs. This raft forms about 1–2 hours after dawn and usually only on bright days with a calm sea. The birds typically arrive at the raft singly and not as a flock. Most are still in winter plumage, with a dark collar. That these birds are associated with the colony, and not wanderers that have just settled there by chance, becomes evident as wind and sea currents drift the birds further from the colony, and they respond by flying back to settle again immediately off the colony site. The raft disperses in the late morning and the individuals fly directly out to sea, disappearing from view beyond the horizon. No kittiwakes are seen near the colony for the rest of the day. However, the raft reforms on subsequent mornings, providing the wind is not strong.

Stage 2 (7–20 January). After the raft of kittiwakes on the sea has formed for several days and has increased in size to over twenty birds, some of the individuals occasionally engage in short outbursts of calling ('kittiwaaking'). On occasion, during these periods of calling, a small group of birds leave the raft and fly silently and directly towards the colony cliffs. The birds do not land, but when about ten metres away they change direction and fly parallel to the cliff, passing close along the cliff area where nesting had occurred in previous years. After one or two passes along the cliff face, they return and resettle within the raft of individuals on the sea. As happened previously, kittiwakes forming the raft disperse before midday.

Stage 3 (15–25 January). The behaviour in Stage 2 is repeated, but during one of the passes parallel to the cliff face, one or more birds alight on nest sites (Figure 3.2). The kittiwakes that alight are extremely nervous, standing with their heads stretched upwards and actively looking around. In addition, the wings, although folded, are held slightly apart from the body, a further sign of insecurity and the readiness to flee. The duration of the visit varies from a few seconds to several minutes, but the birds are never relaxed and are usually silent. Eventually, one bird dives off the site and is immediately followed by all the others in a 'panic' flight. Instead of taking off in a shallow glide which involves the loss of a metre or so in height (as is usual when leaving the nest during the main part of the breeding season), the birds power-dive almost vertically from the site, gaining speed with a few short wing beats, and then pull out into a fast, horizontal flight when a metre or so from the water surface, and return directly to the raft. Panic flights can occur at any time of the breeding season, but are much more frequent at the start and end of colony occupation. During Stage 3, some birds may return to the cliff after a few minutes, but these brief visits invariably result in further panic flights.

Stage 4 (20–27 January). The same routine is followed as in Stage 3, but the birds remain longer in the colony and some birds form pairs (Figure 3.2). The arrival of a potential mate results in repeated bouts of greeting 'kittiwaak' calls, which give the species its vernacular name. By this point, only a few individuals still have a dark collar and most have pure white heads. Panic flights are less frequent, but the daily vacation of the colony in early afternoon is usually preceded by low intensity panic flights.

Stage 5 (February). Kittiwakes arrive during the first hour after sunrise, usually singly but occasionally in groups of three or four birds. They fly directly to the cliff and onto the nest sites, without first forming a raft on the sea. More birds are present in the colony, and the proportion of paired birds and the intensity of vocalisation increase rapidly. The colony is still deserted each day during the early afternoon, often with the synchronous departure of groups of kittiwakes following less intense panic flights, each of which affects only part of the colony. As occurred earlier, the departing birds fly directly out to sea on a steady course, disappearing over the horizon.

Stage 6 (March). Adult kittiwakes arrive at the colony an hour or so before dawn and the colony is occupied throughout daylight hours, with arrivals and departures spread over several hours and the last birds leaving the colony just after sunset. The peak numbers of birds are still present during the morning and there are often over 20% of the birds in pairs, with frequent greeting displays and outbursts of 'kittiwaaking'. Panic flights are infrequent and departures of individuals from the nest site now use the typical shallow dive, which is used by departing birds throughout the rest of the breeding season. The departing birds continue to fly out to sea and disappear beyond the horizon.

Stage 7 (April). This is similar to Stage 6, but there is now little variation in numbers present during the daylight hours. However, all of the adults still depart each evening, although some remain long after sunset, eventually departing in the dark and as late as midnight (Figure 3.2). The colony is still completely deserted during the rest of the night, but some birds have returned by first light. This is the stage when the noise in the colony reaches a peak, with a high proportion (20–25%) of the birds in pairs. There is much displaying, with outbreaks of the characteristic 'kittiwaak' calls. Low intensity nest building starts during this stage.

Stage 8 (late April). This stage does not occur until a few days before the onset of egg laying and is characterised by many individuals and pairs remaining in the colony throughout the night for the first time in that season. Some pairs start intensive nest building during this stage.

Initially, occupation of the colony involves only a few birds and for only a brief period of the day. The progressive changes in the diurnal pattern of occupation are shown in Figure 3.2. Stages 1–8 occurred over a period of about four months at Marsden, but in more recent years at Marsden and at more northern sites, the arrival is later and the stages are passed through within one or two months.

SOCIAL PANIC FLIGHTS

As well as the panic flights seen during the reoccupation and the end of season desertion of the colony, identical mass panic flights at kittiwake colonies are initiated by the appearance near the colony of potential predators

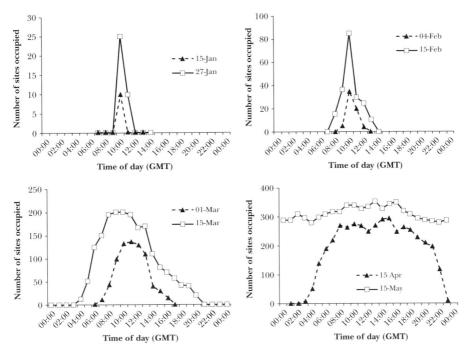

FIG. 3.2 *The diurnal occupation of a kittiwake colony at Marsden, Tyne and Wear on six dates from 15 January to 15 May 1958. Note the short duration of occupation in January and that it is not until March that any sites remain occupied after sunset, but even then the birds did not remain at the colony overnight until late April. The vertical scale changes between the graphs.*

(White-tailed Eagle, Osprey, Peregrine Falcon, Gyr Falcon, Kestrel, feral pigeon, humans climbing the cliffs or a helicopter approaching). All panic flights involve highly synchronised departures, with all or the majority of birds simultaneously pouring off the cliff and flying out to sea. In contrast to the panic flights stimulated by predators, those early in the reoccupation process (and at the end of the breeding season) are spontaneous 'vacuum activities' and are not triggered by an external stimulus, but the manner of departure is exactly the same.

The birds are extremely nervous when they first visit the nesting sites and the urge to flee is readily triggered simply by the diving departure of one or more other individuals. Presumably, this behaviour indicates a conflict between the urge to breed and a fear of returning to the cliffs where potential predators are more likely than over the open sea. When kittiwakes first return to the cliffs, there is a delicate balance between the two conflicting drives, and the dominating one switches over in a short period of time. At first, the urge to return to the cliffs dominates only for a small part of the day, but progressively lengthens as time passes.

WHERE DO THE KITTIWAKES GO WHEN THEY LEAVE THE COLONY EACH DAY?

Birds departing the colonies at Marsden do so on a fixed course. The flight is determined and they do not deviate in direction. Initially, I did not know how far they flew, but some radar observations in Shetland and then off Marsden picked up objects which were presumably birds, flying at about 45km/h departing daily from the coast in spring and then disappearing when about 50–60km from the coast. These observations fit closely with the kittiwake departures and if this link is correct, the departing kittiwakes continue flying for about 90 minutes and presumably feed and sleep well offshore. The radar

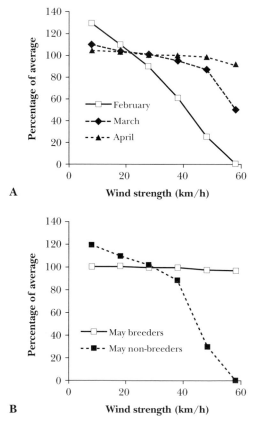

FIG. 3.3   **A.** *The number of breeding kittiwake nest sites occupied in relation to winds of different strengths and in comparison with the average numbers present in that month. Notice how the effects of strong winds on the kittiwakes decreases as the breeding season approaches.* **B.** *The effect of wind strength on the number of sites occupied by breeding and by non-breeding kittiwakes in May. Wind strength has virtually no effect on the numbers of breeding sites occupied, but at the same time the young, non-breeding birds reacted adversely to strong winds, and behaved similarly to the breeding birds in February by appreciably reducing the numbers present on windy days.*

did not pick up birds on the return flight in the morning, but it seems that the return flight to the colony may be more spread out over time, presumably taking individuals about a further 90 minutes to complete.

EFFECTS OF WIND STRENGTH ON PRESENCE AT THE COLONY

During the period of reoccupation, the presence at the colony and the behaviour of kittiwakes are both very much influenced by the wind strength. The early stages of attendance are totally inhibited if the wind speed exceeds 25km/h (16mph). An analysis of the numbers of sites occupied in relation to wind speed shows that the adverse effect of increasing wind speed is linear or curvilinear, but also the effect decreases progressively as the breeding season

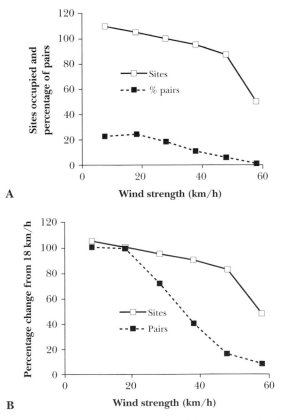

FIG. 3.4   *A comparison of the number of sites occupied (open squares) and the number of sites with pairs (filled squares) at different wind speeds in late March in relation to the wind strength. In B, the values of sites and pairs when the wind speeds were about 10km/h (6.3mph) have been taken as 100%. All counts were taken between 10.00 hrs and noon GMT.*

approaches (Figure 3.3A). In Stage 4 (late January), the effect of an increase from 10 to 20km/h in wind speed reduced the numbers of birds present in the colony by over 50%, whereas in Stage 6 (March) a similar change in wind speed caused only a 20% reduction, and in Stage 8 (April) the reduction is further reduced to only 3%.

It is evident that wind conditions affect the behaviour of kittiwakes and this is not the result of strong winds making landing at nest sites impossible. Most of the returning birds are older individuals and many will have used the same sites in past breeding seasons and encountered strong winds in the past, but these did not deter their landing when they had eggs and young. Support for this interpretation comes from observations made in May on breeding and non-breeding individuals (Figure 3.3B) which show that strong winds deterred the non-breeders attending the colony, while at the same time they had no effect on site occupation by the breeders.

While strong wind reduced the numbers of sites occupied in a colony early in the season, it affected the proportion of pairs present to a much greater extent, with over 20% of sites with pairs when the wind was below 10km/h declining to less than 3% of the occupied sites when the wind exceeded 30km/h (Figure 3.4A and B). Similarly, a wind of approximately 50km/h in March reduced the average number of sites occupied by about 17%, but the number of pairs present in the colony was reduced by 84% (Figure 3.5). This difference suggests that a positive attempt to occupy the nest site by individuals is made at most times, but that the presence of pairs, and so courtship, is mainly restricted to days with low wind speeds. On the few occasions during the return period when strong winds moderated rapidly during the day, there was no increase in the numbers of birds or pairs present at the colony until the following day. The decision to attend or not to attend the colony is apparently made in the early morning, presumably where the birds had spent the night well out to sea. What are the birds which remain at sea doing on windy days? Do they take this as an additional opportunity to feed or do they just sit out the rough conditions until the weather moderates?

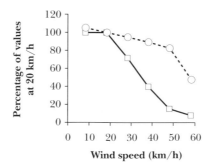

FIG. 3.5   *The effect of wind speed on the percentage of sites occupied (circles and dashed line) and the percentage of pairs (squares and solid line) compared with the average conditions at a wind speed of 20km/h.*

FACTORS AFFECTING THE TIME
OF RETURN OF INDIVIDUALS

Individuals that breed in a particular colony return at different times, and these differences can be measured in weeks and even months, rather than in days. In the same year, some individually marked birds returned in January. Others did not arrive until late April. The date at which different colonies are first visited varies geographically, but it can be different between colonies only a few kilometres apart. Factors which influence the date of return of kittiwakes are considered in detail below.

Age

The effect of age on the time of first return to the colony has been measured for many years, using the colour-ringed birds breeding at North Shields. This showed that there is a major age hierarchy linked with the date of return to a particular colony (Table 3.1 and Figure 3.6). Old kittiwakes are the first to return to the colony and birds which have bred for more than ten years return more than two months earlier than those which are about to breed for the first time. Young individuals which will breed for the first time in the year under consideration enter the colony just prior to the time of laying by the earliest (old) adults, while many of the non-breeding (sub-adult) birds, which will nest in a future year, do not arrive in the colony until there are already chicks in some of the nests, and so they can arrive up to four months after the first individuals returned to the colony. This age-related hierarchy of arrival is also evident among those birds which have bred before (Figure 3.7), with the

TABLE 3.1 *The mean date of first sighting in the colony each year in relation to the number of times breeding will have taken place, including the current breeding season. Data 1960–1967 from North Shields.*

Females	Number of times the individual will have bred including the current year							
	First	Second	Third	Fourth	Fifth	6th or 7th	8th to 10th	Over 10th
Mean days from 1 January	114	61	63	56	50	53	54	42
Mean date	24 Apr	2 Mar	4 Mar	25 Feb	19 Feb	22 Feb	23 Feb	11 Feb
SE	5.0	3.0	3.3	3.8	3.5	2.7	3.2	6.7
Number	19	47	40	39	32	46	29	15

Males	First	Second	Third	Fourth	Fifth	6th or 7th	8th to 10th	Over 10th
Mean days from 1 January	107	60	52	53	41	45	46	38
Mean date	17 Apr	1 Mar	21 Feb	22 Feb	10 Feb	14 Feb	15 Feb	7 Feb
SE	5.0	3.5	3.2	5.1	3.9	3.7	3.8	4.5
Number	27	58	47	34	26	35	23	7

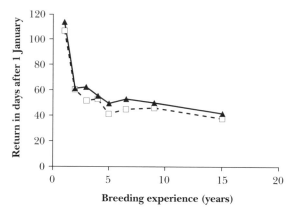

FIG. 3.6 *The date of first return of male (squares) and female (triangles) kittiwakes according to breeding experience including the current year. In general, males return about 6 days earlier than females, while the oldest individuals return about 70 days earlier than those birds about to breed for the first time.*

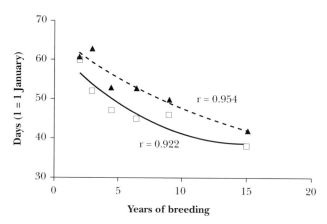

FIG. 3.7 *The average dates of return of kittiwakes (which have bred before) to the North Shields colony in relation to their breeding experience. Males (squares and continuous line) consistently arrive about six days earlier than females (triangles and dashed line). Second-order polynomials have been fitted to the data points as the effect of breeding experience on time of return is non-linear. There is a clear progressive trend in both sexes for older birds to return earlier.*

arrival date becoming progressively earlier with age (increasing number of years of breeding), and produces an average difference of 20 days between birds which bred for the first time last year and birds which had bred at least 12 times. The early return gives an advantage to the older birds, in the choice of both nesting sites and mates. On the other hand, an early return has the disadvantage of restricting where these birds can obtain food, which has to be within easy flying distance of the colony (perhaps 100km or two hours' flying time).

▲ Second year kittiwake. It is the black along the leading edge of the second longest primary (second from the right) that is characteristic of second-year birds; the longest primary has a dark leading edge in kittiwakes of all ages. The rest of the plumage and the yellow bill is that of a full adult (Steve Round).

▼ A recently fledged juvenile kittiwake. The contrast with the photo above shows clearly why early taxonomists thought that kittiwakes in their first-year 'tarrock' plumage were a different species! (John Anderson).

▼ Adult kittiwake showing the bright red gape and mouth, which is only exposed when the yellow bill is opened (Dean Eades).

◀ Kittiwake holding nesting material in its bill. The bird had recently been collecting mud, hence the discoloration of the beak and legs (Michael Osborne).

◀ Pair of kittiwakes reuniting. The bird in flight has already started the moult of the inner primaries, and is being greeted by its mate on the nest. The bird at bottom right has partially closed eyes, and is advertising for a mate (Michael Osborne).

▼ A pair of kittiwakes engaged in mutual display (Robert Barrett).

▲ Kittiwake nest with typical two-egg clutch. The nest is lined with straw, grass, and brown seaweed. The presence of a few feathers is atypical (Becky Coulson).

◄ Colour-ringed kittiwake about to lay an egg, Baltic Flour Mill, Gateshead (Becky Coulson).

▼ Kittiwake with well-grown chick. The adult typically stands between the chick and the edge of the nest, reducing the risk of the chick being pushed off the nest to its death (Michael Osborne).

▲ Adult kittiwake with a six-day-old chick that has lost the egg tooth but retains the grey down that was present when it hatched. The secondaries and primaries are beginning to grow on the wing (Brian Anderson).

◀ A typical kittiwake colony prior to nest-building and laying, showing the close packing of birds and half of the sites occupied by pairs. Jack Rock, Mardsen, Tyne and Wear (Michael Osborne).

◀ The kittiwake colony at Seahouses, Northumberland. Many chicks are close to fully grown, and some have been left unguarded by adults, perhaps at a time of temporary food shortage (Michael Osborne).

▼ Kittiwakes and Common Guillemots; these birds are in competition for breeding sites on Brownsman, Farne Islands. The nests are on hard basalt rock (Michael Osborne).

▲ Kittiwakes nesting on the girders of the Tyne Bridge, Newcastle upon Tyne, despite the sloping surface (Michael Osborne).

▼ Kittiwakes on the artificial 'cliff' at Lowestoft harbour (John Coulson).

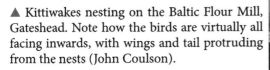

▲ Kittiwakes nesting on the Baltic Flour Mill, Gateshead. Note how the birds are virtually all facing inwards, with wings and tail protruding from the nests (John Coulson).

▼ These kittiwakes have built their nests on spikes placed to deter them on the Guildhall, Newcastle upon Tyne (Michael Osborne).

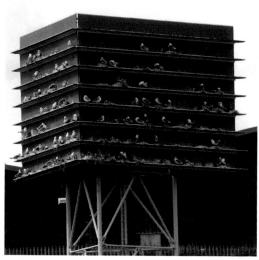

▲ The tower put up at Gateshead, Tyne and Wear, to replace nesting sites on the Baltic Flour Mill (Michael Osborne).

▲ Kittiwakes nesting on a boulder beach, Hirsholmene, Denmark (John Coulson).

▼ A young White-tailed Sea Eagle flying close to a kittiwake colony, causing much disturbance to the breeding birds. The kittiwakes are alarmed but are not attempting to attack the eagle. Such disturbance has become more extensive in recent years, particularly in northern Norway and Russia, and it seems to have led to reduced breeding success in the kittiwakes, some caused by other avian predators taking eggs and chicks during the disturbance (Hans Ueli Grütter).

▲ A one-year old kittiwake in July, partially through its first moult. The dark nape is still retained, but it is less marked than in the juvenile, while some of the black tips to the tail feathers remain. Much (but not all) of the bill has become yellow; the three outer primaries of the juvenile plumage remain, but the black component has faded to brown and so contrasts with the black tips to the new and still-growing 5th and 6th primaries (Maarten van Kleinwee).

▼ Adult in moult, July, Netherlands. Note the completely black tips to the outer five primaries, and the absence of white 'mirrors'. The outer primaries are about 11 months old but the black tips have not degraded to dark brown (see above). The inner two primaries have been replaced, two more have been dropped but have not yet been replaced and the rest are old feathers from the previous year (Maarten van Kleinwee).

▲ A juvenile kittiwake carrying out wing exercises, but without lifting off from the cliff-edge nest site. Note the similar length of the four outer primary feathers. These feathers give the wing of the young bird a shorter, blunter appearance than that of the adult (Michael Osborne).

▲ A pair of Red-legged Kittiwakes. Note the larger eyes and shorter bill (Morten Jørgensen).

▼ Adult Red-legged Kittiwake next to adult Black-legged Kittiwake (Morten Jørgensen).

The mean date of first return by males is about six days earlier than for females, but this may not be of major biological significance, other than to indicate that males occupy the sites first and initially spend more time at the colony, which contributes to the males being identified earlier than females. However the rapid formation of pairs when the colony is first occupied each year indicates that both sexes tend to return at very similar times and this probably contributes to the re-establishment of many, but not all, pairs that existed in the previous year.

COMPARISON BETWEEN COLONIES
IN THE SAME IMMEDIATE AREA

During studies at Marsden in the 1950s, it was evident that there was considerable variation in the timing of reoccupation, despite the colonies being within the same 3km stretch of the coast. In Figure 3.8, the colonies have been grouped into those on a stack, called Marsden Rock, which contained 1,407 nests in 1954 and was the original nesting site in the area, the mainland colonies formed later with between 111 and 197 nests each, and three new colonies, formed within the previous three years and with 13–20 nests each. It is immediately evident that the reoccupation and build-up of numbers differed markedly between the three categories. Both the first occupation and the date by which half of the nest sites used for breeding in that year were occupied were markedly different. The timing of first occupation of the 'Rock' and the 'Young' colonies differed by 60 days and the dates of 50% occupancy were 40 days apart. Even by mid-April (and within 14 days of the first eggs to be laid at Marsden), only 75% of the nest sites were occupied in the Young colonies but 97% on the Rock. Throughout, the Mainland colonies were intermediate in the build-up of occupied sites.

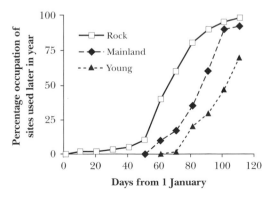

FIG. 3.8   *The return and build-up of occupied sites at three groups of kittiwake colonies at Marsden in 1954. The number of nests built has been used as the basis of the percentages. Similar differences existed in 1953, 1955 and 1956. Based on Coulson & White (1956).*

The differences can be attributed to two causes:

1. The largest number of birds nesting on the Rock and the smallest in the Young colonies resulted in a considerable difference before the raft of returning birds formed on the sea was large enough to reach numbers which stimulated visits to the cliffs.
2. The Rock had, presumably, the largest proportion of older, experienced breeders which return first, while these were lacking in the Young colonies. The Young colonies each increased in size more rapidly than the others and so had the highest proportion of new breeding birds, and this must have contributed to the smaller number of occupied sites even in mid-April.

While I believe that these two effects contributed to the differences, they do not entirely explain all of the differences, and a third effect, the greatest density of nesting kittiwakes on the Rock, probably also made a difference. This would imply that the nest density in the previous year influenced the time of return this year and consideration of this effect is returned to later. Even in the same area, the pattern of return can vary markedly between colonies, depending on their size and density and the proportion of birds of different ages they contain.

Past breeding success

Individual kittiwakes which nested successfully in the previous year tend to return to the colony earlier and at a more consistent date in the current year than those of the same age but which failed to produce young last year. This affects only a minority of the birds in the colony in any year when the breeding success is high (as it was in the 1950s at Marsden). Further, the difference is only a few days and so is unlikely to affect the timing of the reoccupation and return to the colony to any appreciable effect or to explain differences between colonies which had had similar breeding success in the previous year. This aspect is considered in more detail in Chapter 8.

Changes in the time of reoccupation, 1952–2006

Occasionally, local bird reports comment that 'the kittiwake colony was occupied a few days later this year than last year', but observers have not noticed that the annual reoccupation of the colony at the start of the 21st century is now weeks and even months later than it was thirty years earlier.

After the later but advancing return dates as the new colony at North Shields grew in the 1950s, the timing of return at North Shields and Marsden thereafter tended to be similar and I have collated the date of the first reoccupation of these colonies for 54 consecutive years (Figure 3.9). In the 1950s, the first sites were occupied at North Shields during early February (35 to 50 days after 1 January) but by the 1970s, the first birds were back on sites in January (and in one year in mid-December). By 1990, the first return was in late February, with progressively later arrivals from 1977 to 1990. In recent years, the first returns to Marsden have become even later, with no birds on nest sites until the second half of March since 1999. The early return and reoccupation in the 1950s was not evident in other colonies in eastern England and Scotland, where the first reoccupation was typically at the end of February. However, the later arrival in recent years has occurred in the other colonies in north-east England and so this progressive later return is not specific to the Tyne area, nor was it related to the size of the colony.

The earlier arrivals from 1960 to 1978 cannot be attributed to changes in the size of the colony, as the numbers of nests were similar throughout this period, nor is it attributable to any other characteristics of the colony during that period. Similarly, there is no suitable explanation in the age composition of the birds to account for the progressive later annual arrivals from 1990 to 2008. The late arrivals during the period 1990 to 2006 have also been reported in other areas below the Arctic Circle in western Europe, with kittiwakes not arriving until late March in most years. So why have kittiwakes changed the time of return to the colonies? The cause of the change is unlikely to be at or in the vicinities of the colonies, because kittiwakes return to the coastal waters and to the colonies at very much the same time. They are not spending much time in waters close to the colonies before reoccupying nesting sites. It appears that they are remaining in the winter area longer in recent years and

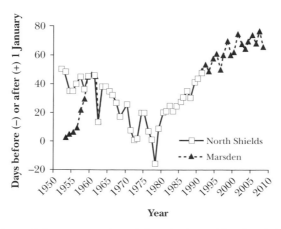

FIG. 3.9 *The dates of first reoccupation of the colonies at Marsden (triangles) and North Shields (squares), 1952–2008. Note the single reoccupation in December and the progressive later reoccupation since about 1978. Data for Marsden 1985–1990 were the same as those at North Shields and are not shown.*

the cause of the change would appear to lie in the conditions in the oceans during the winter and not to factors affecting the coastal waters around colonies, because the birds were already late when they arrived in the coastal areas. Little is known about food availability in pelagic areas, and how this affects the timing of the birds returning to inshore waters and the colonies.

How is the time of return to the colony determined?

In the 1950s and 1960s, kittiwakes nesting in colonies in the vicinity of the River Tyne usually made their first return to their colonies by the end of January, 4–6 weeks earlier than colonies elsewhere in north-east England. The reason for this difference is not clear, but by 1980, the first birds in all of the colonies were arriving in mid-February or even later. Ten years later, the start of reoccupation had been delayed until the last few days of February, while in the early 2000s the first arrival was delayed until the first or second week of March. Not only were colonies occupied later, but the earlier return of kittiwakes into coastal waters was similarly delayed. These progressive delays in the return are, of course, the opposite of what has been observed in many plants and animals in Britain, where flowering, bud-burst, emergence and breeding have become earlier in recent years and linked to progressively warmer temperatures. To put it simply, the kittiwakes were leaving their oceanic distribution progressively later and so could not be reacting to climatic changes in the vicinity of the colonies. But here we draw a blank, because there is little information about the biotic conditions of the open oceans. Have conditions in the North Atlantic changed and, if so, then in what ways?

An aspect of the reoccupation of kittiwake colonies that raises many interesting biological questions is the differences in the time of return of nesting kittiwakes to colonies only a few kilometres apart, and even in different parts of a single colony. In a detailed study at Marsden, Tyne and Wear, the differences in the time of occupation of different colonies were consistent year after year. How did these differences arise and when and how were they produced? It is hard to believe that birds nesting only a few hundred metres apart retain similar spatial relationships outside the breeding season, so differences in the timing of the reoccupation between these colonies and groups cannot be attributed to events in their wintering areas. They have to be produced and then retained from the previous breeding season. In general, birds at high-density colonies breed earlier and then return earlier to the colony in the next spring. Further, because kittiwakes are nomadic in their winter distribution, and encounter fluctuating day length and temperatures as they move around the ocean, it is difficult to propose that these might be the only or main stimuli triggering the annual return to coastal waters and colonies. Similar problems exist in explaining how land birds start migration when wintering near the equator, because here photoperiod does not change seasonally and so cannot give a clue about the changing seasons.

There is now much evidence from research in Germany and elsewhere that such bird species use an 'internal clock' (an endogenous rhythm) which is primed in the previous year and later triggers gonad development and the onset of return migration without an external stimulus in the wintering area being involved. It is likely that such a system also exists in the kittiwake. If it does, the setting of the 'internal clock' would seem to occur in the previous breeding season, and it presumably starts earlier in older and successful breeding individuals, those which nest at a high density, and later in young, immature birds. It is possible that feeding conditions in the winter when kitti-wakes are oceanic may modify this timing, accounting for the later return in recent years. All of this, as applied to the kittiwake, is but a hypothesis, but it is a feasible explanation for the differences in the timing of return of breeding (and non-breeding) kittiwakes which have been observed and is given further support by the earlier return of the birds nesting at the highest densities within one colony.

CHAPTER 4

Pre-breeding, nesting and nest building

ONCE numbers of kittiwakes have returned to the colony, pairs are formed rapidly and the colony becomes a place with frequent outbursts of loud calling and activity. In many cases, the same pairs that bred together in the previous year are reformed, and the effects of this fidelity are examined in more detail in a later chapter. Males are first to occupy nest sites, and from there they advertise and call in an attempt to attract females flying past, be it their mate from the previous year or a new partner. If the mate from the previous year has died, or even if her return has been delayed, the male does not usually wait very long, but soon attracts a new partner, often a female which nested nearby last year or an individual which has not bred before.

There is no evidence that pairs stay together in the winter, and they have to return to the colony at similar times to remain together in the next breeding season. Members of pairs which bred successfully last year tend to return at more similar dates than those that failed, resulting in a higher proportion of previously successful pairs reuniting for the next breeding season. Again, mate retention is more frequent among the older birds, possibly because there are fewer others to choose from early in the season.

Initially, both members of the pair spend much time together at the nest site, which presumably strengthens the pair bond, and there is much mutual displaying and calling, with bowing and 'kittiwaak' calls, whenever the pair reunites at the nest site. Following the initial reoccupation of the colony each year, there is a progressive and rapid increase in the proportion of occupied sites at any one time that are occupied by a pair, and this reaches an average of about 23% (Figure 4.1) and occasionally up to 30% of the occupied sites during calm, spring-like weather. Not only is the mutual calling noisy, but the frequency and volume are increased by an infectious spread of calling to neighbouring pairs. In contrast, a solitary bird of either sex on a nest site usually shows little or no response to the calling of neighbouring pairs. The frequency of calling is increased further by one of the pair departing, briefly flying around the colony and then returning to its mate within a minute, which initiates yet another bout of greeting and calling. Sometimes these 'circuits and bumps' (in aviation parlance) are repeated many times, with each landing stimulating courtship between the pair and also triggering neighbouring pairs into more displaying and 'kittiwaaking'.

The proportion of time spent together as a pair declines as egg laying approaches (Figure 4.1) and a shift system develops, with one member of the pair being present at the nest site while the other is away, a pattern which is a precursor to the shift system evident during incubation, and the nest site is rarely left unoccupied. This period of decline in the percentage of time together also coincides with the start of intensive nest building. However, during this pre-egg stage, which lasts several weeks, day length is increasing rapidly and so the total amount of time pairs spend together does not decline quite as rapidly as the percentage of time spent together.

One of the characteristics of colonial nesting in the kittiwake is that pairs retain ownership of the nest site during the breeding season by keeping it

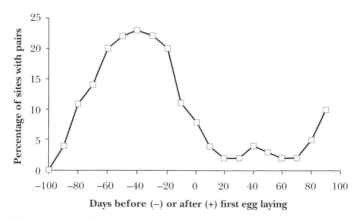

FIG. 4.1   *The percentage of occupied kittiwake nesting sites with pairs during daylight, in relation to the date of egg laying. Based on counts at Marsden 1953–1960 and North Shields 1955–1970, and on 2,678 pair-years.*

occupied (Figure 4.2) and so deterring takeover of the site by prospecting individuals. In Figure 4.2, the high proportion of the daylight hours during which the site was unoccupied early in the breeding season is primarily caused by the arrival of all of the birds in the colony after sunrise and their departure before sunset, and the percentages are reduced by over half if this is taken into account. The proverb 'Possession is nine-tenths of the law' is highly appropriate to kittiwakes.

Much of the time that pairs are together is spent in courtship and pair bonding, but when intensive nest building begins (about 10 days before laying) and the time of laying approaches, the behaviour of the pairs develops into a shift system, with less time spent together, a pattern that becomes even more pronounced during incubation and the care of the chicks. This behaviour increases the amount of daytime potentially available for feeding. In early April, on average an individual adult was present at the colony for about 58% of the daylight hours each day, leaving 42% of the daylight for feeding. Both leave the colony at night, but I suspect that kittiwakes at this time of year do not feed extensively in the dark, and spend much of this period resting on the sea in areas where they can feed at dawn. On the other hand, birds arriving at the colony after dawn sometimes regurgitate recently caught fish, and perhaps intensive feeding takes place during the first light of day and before returning to the colony.

Since day length progressively increases in the spring, the actual amount of daylight available to birds for both feeding and breeding activities at the colony increases and the time allocated to feeding is probably comparable to that available at similar latitudes in the winter (when kittiwakes are pelagic) and there are no other competing factors apart from avoiding severe weather.

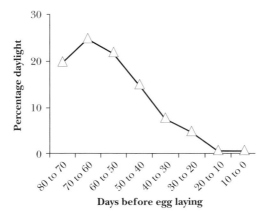

FIG. 4.2   *The percentage of daylight hours during which kittiwake nest sites remained unoccupied in relation to the number of days before egg laying occurred. Based on data from detailed records obtained for 13 pairs by time-lapse photography or radioactive marked pairs.*

EFFECT OF COURTSHIP

The outbursts of courtship greetings are probably crucial in bringing the female into mating and egg-laying condition. Studies made with Fiona Dixon showed that the frequency of 'kittiwaak' calling in a pair was related to the number of other pairs in close proximity, usually within a two-metre radius. As a result, those nesting at high density were stimulated by more neighbours into calling and displaying, since there were more pairs near to each focal nest site (Figures 4.3 and 4.4). If it is assumed that the focal pair receive increased stimulation when each of their neighbouring pairs reunite and display, then a pair at a high density site receive almost twice (i.e., the sum of

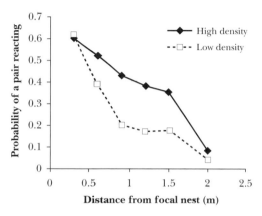

FIG. 4.3   *The probability that pairs of kittiwakes will respond to an arrival ceremony by the focal pair in relation to their distance apart.*

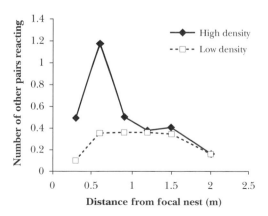

FIG. 4.4   *The number of pairs of kittiwakes responding to the arrival ceremony in relation to the distance from the focal nest for parts of a colony with a high and a low density of nests. The total of the numbers at the two densities indicates that a pair at the high density accumulate almost twice the amount of stimulation from neighbouring pairs than a pair at the low density.*

the points in Figure 4.4) the stimulation as those at low density. This extra stimulation results in earlier egg laying among kittiwake pairs nesting in the high-density areas of a colony as they often have near pairs on all sides of them. In contrast, edge-nesting kittiwakes receive their stimulation from only one direction and the density of neighbouring pairs is typically low at the edge. As a result, the edge pairs accumulate stimulation at a slower rate, resulting in later laying (and the production of more single-egg clutches).

Applying the importance of social stimulation further, it becomes evident that isolated pairs of kittiwakes do not have access to the necessary stimulation from neighbouring pairs, and fail to breed. As a result, kittiwakes are obligatorily colonial. The 'rule' which applies to kittiwakes, and probably many other colonial species, is that 'you have to be in a group or you cannot breed'. Such a mechanism ensures that coloniality persists, and prevents individual pairs separating from the group and breeding in isolation. It also explains the advantage of an early return to the colony by kittiwakes, as they receive stimulation earlier, and emphasises the importance of the severe competition arising from the need to obtain a site with many neighbours in the restricted space of the colony.

A search of the literature reveals that there are very few seabird species that exhibit both solitary and colonial nesting. Isolated nesting by pairs is known to occur in the Common Tern, Common Gull, Great Black-backed Gull and older Herring Gulls (but not usually in young breeders), but is exceptional and does not occur in most other colonial species.

NEST BUILDING

Nest building by kittiwakes is both a solitary and a communal activity. It starts as a low intensity activity about three weeks before the first eggs are laid, and at that time it is a highly variable and unpredictable process, with one bird occasionally bringing in a piece of nesting material. Ester Cullen (1957) suggested that the onset of nest building was often associated with recent rainfall because mud, which is included in the foundation of the nest, becomes more readily available at that time. During and after rain, water seepage often produces mud at run-off sites at localised points on sea cliffs and at the edge of temporary freshwater pools near the cliff tops and this is the only time kittiwakes visit these sites during the year. Green or brown seaweeds are gathered, and sometimes grass and short lengths of rope and twine are also collected. The mixture is deliberately trampled by the adults onto the surface of the nesting ledge. As this mixture dries, it becomes firmly attached to the rock ledge and forms a base for the rest of the nest.

At times, large numbers of kittiwakes are attracted to these restricted sources of mud, and this results in a stream of kittiwakes flying between their nest sites and the mud, so it becomes a highly social activity that can involve tens and even hundreds of kittiwakes walking around together collecting mouthfuls of mud. The mud is carried to the nest site in the bill and deposited there by vigorous bill shaking. This almost haphazard deposition of mud was a major problem at the study colony at North Shields (where nesting occurred on windowsills of a warehouse) because it soon made the glass on the windows opaque. It also had the advantage of effectively indicating when nest building started! Birds at the nest sites shake their bills to remove the last of the mud from their mouths and this forms a crescent of mud on the vertical surface of the rock behind the nest site at the height of the bird's head. On dark rock, the mud crescent is not obvious.

Having completed the nest platform, more grasses and seaweeds are collected from the tideline and from the cliff edges, often from or near the same sites used earlier to collect mud. The collection of materials by hundreds of kittiwakes at a small number of locations can have a marked effect on the vegetation, where bare patches of ground are produced locally, while small pools increase considerably in size as materials are removed from their edges.

The upper part of the nest is built on top of the mud-based foundation, using drier material such as grass, seaweed, pieces of rope, string, pieces of fishing line and other items picked up at the same sites from which mud was collected, along the tideline, from intertidal rocks and from the sea surface. This dry material is added to the base and shaped into a bowl by a rotating action of the breast driven by a pushing action of the feet. Material is usually collected within two kilometres of the colony, but on a few occasions, I have seen adult kittiwakes over 10km away carrying nesting material and flying in a determined fashion towards the nearest colony.

In one year, a mixture of straw and manure was spread on a cliff-top field in April near the kittiwake colony at Marsden. This was rapidly utilised by kittiwakes and lengths of straw were incorporated into the nests, with strands of straw dangling up to 40cm from almost every nest, giving the colony a bizarre look and an atypical golden sheen.

While established breeders may spend up to three weeks in active nest building, late-arriving pairs may spend only about 3–4 days on producing a nest. These latter nests are much less substantial and often lack both an appreciable mud base and a deep cup to retain the eggs.

Occasionally, one- and two-year-old kittiwakes collect vegetation along with adults. They collect the material satisfactorily, but do not have a nest site to bring it to and wander around with it in their bill for some time until it is finally dropped. In July, recently fledged chicks briefly picked up nest material from the shore, along with two- and three-year-old non-breeders which were still nest building on newly obtained sites, although with no hope of breeding in that year.

In some colonies, the nests are washed off by winter storms and have to be completely rebuilt each year. In others, nest remnants survive from the previous year, albeit with the cup no longer present, having been flattened by the activity of last year's chicks and then by the winter weather. In these cases, new nests are built on top of the previous foundations and while less material is collected, the nests on these sites tend to be taller than those built on the rock ledge. On Staple Island of the Farne Islands group, Northumberland, a nest was built at the base of a small, narrow, vertical cleft in the cliff, which gave the nest material support and protection on three sides. The year-by-year accumulation of nest material in the cleft has resulted in this nest now being over a metre tall, representing over 15 years of accumulations. Because of the annual accumulation of nest material on some of the ledges at the North Shields warehouse, some of the kittiwake nests almost reached the top of the small windows. The most extensive of these were removed during the winter, but this had no effect on when egg laying occurred at these sites in the next breeding season, although they resulted in a greater amount of material being brought in during nest building. In all cases, nest building ceased a few days before the first egg was laid.

CHAPTER 5

Eggs, clutch size and incubation

CLUTCH SIZE

THE Red-legged Kittiwake usually lays a single egg clutch, with an average of 0.5–2% of pairs producing two-egg clutches, although in some years no two-egg clutches were found (Hunt *et al.* 1981, Johnson & Baker 1985, Lloyd 1985, Byrd & Williams 1993). There is a suggestion that the clutch size used to be larger (Byrd & Williams 1993) but this may only reflect that the average clutch size varies between years.

The Black-legged Kittiwake usually lays a clutch of two eggs, with one- and three-egg clutches also occurring. As a result, the typical average clutch size is usually just below two eggs per nest, as single-egg clutches tend to be slightly more frequent in most colonies than those composed of three eggs. However, the data collected at North Shields, north-east England, over the period 1954–1990 averaged 2.03 eggs and varied very little from year to year. In contrast, the clutch size of the kittiwake in the Arctic often varies markedly from year to year. This is well illustrated by the data collected by Belopolskii (1957) on the Murmansk coast of Arctic Russia. In four successive years, the average clutch sizes were 2.33, 1.53, 1.74 and 2.03 eggs, with an overall average of 1.91 eggs.

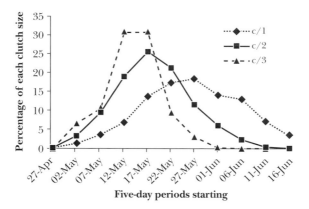

FIG. 5.1 *The percentage of the total of clutches of one, two or three eggs laid by kittiwakes at North Shields 1954–1990 in each 5-day period of the breeding season centred on the date shown. Based on 276 single-egg clutches, 2,068 two-egg clutches, and 315 three-egg clutches. These differences lead to the clutch size declining with date.*

TABLE 5.1 *The proportion of female kittiwakes replacing a lost clutch in relation to their breeding experience.*

	Breeding experience of female (years)			
	1	*2*	*3 and 4*	*Over 4*
Number lost	23	13	10	14
Number replaced	0	3	6	10
Percent replaced	0%	23%	60%	71%

The time of laying different clutch sizes varied consistently with date in individual colonies. Clutches of three eggs were confined to the first half of the breeding season, while single-egg clutches peaked in the latter part of the breeding season and this is illustrated for the North Shields colony in Figure 5.1. This pattern gives rise to a declining clutch size with date. More detailed consideration of the causes of change in clutch size during the breeding season is given in Chapter 8.

RELAYING

Both species of kittiwakes are able to relay if the first clutch is lost, but appreciably less than half do so.

In northern Norway, Barrett (1978) reported that in 22% (n = 253) of pairs which lost their clutch, these were replaced. First-time breeding females rarely, if ever, relaid and only a proportion of older breeding females relaid after losing their clutch (Table 5.1). In contrast, many but not all older females which lost clutches replaced them and the proportion doing so was highest in those which initially laid early in the season and then soon lost their eggs. Cases in which only one egg was lost from clutches of two and three resulted in the continuation of incubation of the remaining egg(s) and the lost egg was not replaced.

At North Shields, the total loss of a clutch was infrequent and caused by several different factors. In some cases, eggs disappeared from a site where human access was impossible and no avian egg predators were present, and they were probably lost during a dispute for ownership of the site. In other cases, the loss was associated with gales and the accompanying strong up-draughts and gusts which occasionally rolled eggs out of the nest. This was a risk only for eggs in nests with shallow cups, which were usually built by young birds. Human interference was rare, but when it occurred it involved only a few clutches in any year.

Wooller (1973) undertook experimental clutch removals and transfers of the eggs to create larger clutches in other nests on the Farne Islands. These involved pairs of unknown breeding experience, but have been included in the data presented below. In cases where clutches were replaced, there was an appreciable time-lag before they were replaced, while the chances of replace-ment decreased with the length of incubation of the initial clutch before the loss (Table 5.2). Clutches which had been incubated for more than 16 days were rarely replaced.

On average, the first egg of the replacement clutch was laid 12 days following the loss of the original clutch, and Barrett (1978) reported a 13-day delay. This delay increased with the length of time the first clutch had been incubated before the loss. Clutches lost 1–5 days after the clutch was completed resulted in the first replacement egg being laid an average of 11 days later, while those lost after more than 10 days of incubation took 14 days before relaying started. Barrett (1978) did not find any relaying if kittiwakes lost their first clutch more than 10 days after its completion. Presumably, these differences were influenced by the more extensive regression of the ovaries as the initial clutch incubation continued, and so a longer time was required for their redevelopment and after about 10 days redevelopment was not possible.

On average, the number of eggs in the replacement clutch was the same as or smaller than in the initial clutch and the volume of each egg was less. No clutches of three eggs were laid as replacements, with clutches of one egg

TABLE 5.2 *The proportion of kittiwake clutches that were lost and replaced in relation to the stage of incubation when the first clutch was lost.*

	Days eggs lost after start of incubation			
	1–5	*6–10*	*11–15*	*16–25*
Total clutches lost	55	39	17	15
Total replaced	38	23	5	1
Percent replaced	69%	59%	29%	7%

more frequent, perhaps reflecting the later date of laying as they are in line with the smaller clutches and egg size in first clutches as the date of laying became later. This decline may not reflect a cost and stress on the females producing a second clutch, as suggested by some (Gasparini *et al.* 2006, but see Monaghan *et al.* 1998), but is simply responding to the normal seasonal pattern of clutch size in kittiwakes.

No clutches lost in north-east England after 24 May were ever replaced. It would appear that relaying (as well as late first clutches) was inhibited by an environmental factor which prevented the laying of first clutches after 9 June. The timing of breeding was consistently later in northern Norway (mean egg date over three years was 1 June; Barrett 1978) and the cut-off date inhibiting laying and relaying there appeared to be about 22 June (data in Barrett 1978). In no cases was a replacement clutch the last to be laid in the colony in any breeding season in north-east England, although Barrett (1978) did record this in northern Norway. Late breeders did not relay if they lost their clutches, presumably because of the effect of the cut-off date. However, the existence of a cut-off date was not the sole reason why first-time breeders did not relay, because early layers which lost their eggs at North Shields before the middle of May did not relay.

DATE OF EGG LAYING

In general the date of breeding in kittiwakes becomes later in more northern areas and, in general, is related to sea temperature, as well as being influenced by the date of reoccupation of the colony.

The earliest breeding occurred at North Shields, when in a few years the first eggs were laid in the last week of April, but typically laying began in May and ended about 10 June. On average, the laying date was 18 May and there was little variation between years. Breeding elsewhere in Britain tended to start about a week later, but ended at the same time. Breeding in southern Norway was not much later, but in northern Norway, Barrett (1978) found the whole laying period was about 11 days later than at North Shields, with a mean laying date of 1 June, but with more variation between years. On the east Murmansk coast of Russia, Belopolskii (1957) found the onset of laying varied considerably from year to year, differing by 18 days between years, but in general these were not different from the dates for northern Norway. However, in Novaya Zemlya, the start of laying was much later and averaged 7 July, with a peak about 25 July.

In the Pacific Ocean, egg laying by Black-legged Kittiwakes is later than in much of the Atlantic, reflecting the fact that they breed further north and in colder waters. On Middleton Island, the peak of laying was about 7 June (Gill 1999) and a similar date applied to Cape Thompson, but in Cook Inlet, the average date of laying was about 3 July and laying is even later in some other Alaskan colonies. Relatively few studies have been made on the timing of egg laying in the Pribilof Islands and most data have been indirectly obtained from hatching dates. Baird (1994) gives very little information on the date of egg laying of kittiwakes breeding in the USA and Canada, and the suggestion in his Figure 5 that the start of egg laying can occur throughout April would appear to be erroneous.

KITTIWAKE EGGS

The eggs of the kittiwake differ from those of most other gulls in that they have a paler background and show close similarities with the eggs of the Slender-billed Gull and Sandwich Tern. Like all other gulls, the eggs are marked with blotches of darker colour, mainly in various shades of brown. Several people have commented that kittiwakes appear to lay relatively larger eggs for their size than other similar sized gulls, and I also gained this impression, believing that this may be a compensation for the smaller clutch size. However, Figure 5.2 shows that the volume of an average kittiwake egg lies precisely on the curved relationship of egg volume plotted against average female weight for a series of Palearctic gulls and terns, and this curve is not influenced by the typical clutch size laid by the species; those species laying smaller clutches do not produce larger eggs. The impression that kittiwakes lay relatively large eggs appears to be an illusion produced by the light background colouring of the shells.

Incidentally, note that in Figure 5.2, three of the skua species produce smaller eggs than would be expected from the weight of the adult females. Is this difference related to female skuas being larger than males, in contrast to

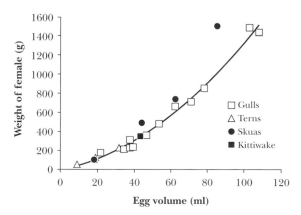

FIG. 5.2   *The points and trend line of the weight of an average female and the egg volume for a number of Palearctic gull, tern and skua species. The kittiwake (filled square) lies on the trend line. There is no indication that the kittiwake, Ivory Gull, Sabine's Gull, Sandwich Tern and Arctic Tern, all of which typically lay two-egg clutches, deviate from the general trend, although three of the four species of skua lay smaller eggs than predicted from the size of the female.*

females being smaller than males in most gulls and terns? If the average size of both sexes is used, the difference of the three smaller skuas from other larids virtually disappears, although the Great Skua remains an exception.

Comparisons made within most major bird taxonomic groups show that egg size becomes smaller in relation to adult mass as adult size increases (Lack 1968). This is also true for gulls and terns (Figure 5.3). Here again, the kittiwake lies on the overall trend line, with each egg being about 14% of the adult female weight. In contrast, each egg of the much larger Great Black-backed Gull is only about 8% of the female's body weight.

EGG VOLUME

While all kittiwake eggs look superficially similar, they vary extensively in volume, although this variation is not immediately evident on inspection because a modest change in the breadth of the egg has an appreciable influence on its volume and weight.

Kittiwakes in areas where they are structurally larger (typically further north) lay larger eggs, and this has already been mentioned and described in Chapter 1. Even within a colony, there is a slight tendency for larger females to lay slightly larger eggs, but this effect is small and was only convincingly detected after very large samples of females and their eggs were considered.

The volume of eggs varies considerably within a single colony. The largest egg in the North Shields colony had a volume of 53.5ml, while the smallest was only 33.0ml. Could the variation be a measure of the condition or quality of the individual females laying the eggs? If this was so, it would be a valuable tool to assist in evaluating variation in the quality of individual females.

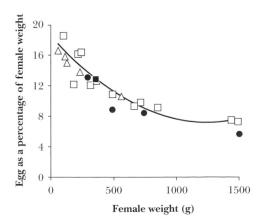

FIG. 5.3 *The relationship between body weight of the female and the mass of an egg as a percentage of the adult female mass for Palearctic gulls, skuas and terns. Symbols as in Figure 5.2. The kittiwake (filled square) lies on the general trend line. Note that the four skua species (filled circles) lie below the trend line for gulls and terns.*

TABLE 5.3 *The average volume of kittiwake eggs at North Shields according to clutch size and order of laying. C/ indicates clutch size and letters a–c indicate order of laying.*

Clutch size and order	Sample size	Volume (ml)	SD	SE
C/1	167	41.43	4.42	0.34
C/2a	1219	43.71	3.33	0.10
C/2b	1219	42.02	3.34	0.10
C/3a	178	43.80	2.91	0.22
C/3b	178	43.49	3.17	0.24
C/3c	178	41.40	2.93	0.33

The volumes of most eggs laid in the North Shields colony were determined in many years of the study and, while there were very small and non-meaningful variations between years, there was no long-term change in the average egg size. However, when the size of eggs in single-egg clutches as compared with clutches of two and three eggs differences existed (Table 5.3). On average, single eggs were smaller than either egg in clutches of two or the first two eggs laid in three-egg clutches. The third egg in a three-egg clutch was smaller than either of the first two eggs, and one of these was invariably the smallest egg laid each year in the colony. The first egg in a clutch of two eggs was larger than the second egg by an average of 1.69ml (4% larger) while the first and second eggs in a three-egg clutch were similar in size. Note that the last egg in a clutch (including the egg in a single-egg clutch) was small and this suggests that the transfer of material from the female into the egg was being gradually turned off as the last egg was being produced.

There is a positive correlation between egg size and the chances of it producing a fledged chick. Single-egg clutches and the last egg in a three-egg clutch had the lowest chance of producing a fledged chick. In two-egg clutches, the larger first-laid egg was more successful than the second egg. But how does egg size affect the level of breeding success? Is it a direct effect or an indirect one, working through the condition and age of the parents?

One approach to this question is to consider the hatching success of eggs of different sizes in single-egg clutches, where there can be no confusion between linking the individual egg and any chick produced from it. This is examined in Figure 5.4 for 137 single-egg clutches, and a clear and progressive correlation exists between size and successful hatching, so that eggs over 43ml had twice the chance of hatching compared with those smaller than 36ml. This is not an unexpected result, because chicks weighed at hatching were heavier when they came from larger eggs with greater yolk reserves which allowed them more time to persuade their parents to start feeding them (see weight at hatching in Chapter 6).

The size of eggs laid by kittiwakes at North Shields tended to decrease as the laying season progressed and this was most pronounced in single-egg clutches (Figure 5.5). However, two-egg clutches did not produce eggs as

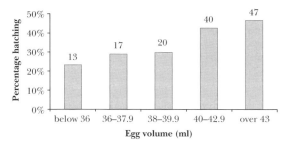

FIG. 5.4   *The hatching success of one-egg clutches laid by kittiwakes in relation to egg volume. The sample sizes is given above each column.*

small as some of those in one-egg clutches, and only 4% of two-egg clutches contained an egg which was less than 38ml, and less than 0.7% of clutches had both eggs below this size. Of the eggs that were less than 38ml, only 32% hatched, compared with 78% of the larger eggs in two-egg clutches.

In the case of three-egg clutches (based on 245 clutches), the third egg averaged only 41.4ml (5.2% smaller than either of the first two eggs), and only 41% of these hatched compared with 72% of the larger first and second eggs in the same clutches. So small eggs had a lower hatching success, but not as low in clutches of two and three eggs as would have been predicted from the data relating to single-egg clutches. This suggests that other factors are involved, for example, larger eggs and clutches tend to be laid by older females, and both age and experience probably play a role in hatching success.

Callum Thomas (1983) attempted to resolve these two possibilities by exchanging eggs between kittiwakes which laid large and small eggs, both produced within the same narrow time period. His results, using two-egg clutches, confirmed that both egg size and experience of the adults played a role in determining hatching success.

In all three clutch sizes laid by kittiwakes, egg volume was larger in early

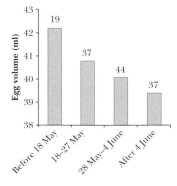

FIG. 5.5   *The average volume of single-egg clutches laid by kittiwakes in relation to the date of laying. The sample size in each time period is given above each column.*

breeding individuals and this effect was most pronounced in single-egg clutches (Figure 5.5). It seems that egg size, even more so than clutch size, is an indicator of the quality of the female and possibly also of the male partner.

Recently, Moncada (2008) measured the chemical composition of kittiwake eggs and gave details of lipid and yolk composition in a range of eggs. The larger egg in clutches of two contained more lipid and yolk, but the amount was not in a constant proportion to the egg size. Reading from the graph of yolk plotted against egg size, large eggs (50ml) contained about 6.1g of yolk and this decreased progressively to 5.0g in eggs only 38ml in size. One possible implication of this is that, providing the structural size of the chick did not vary markedly with egg size, then a newly hatched chick from the larger egg probably retained more lipid and yolk stored within its body and this was evident from external examination through the skin of newly hatched chicks. This is similar to the situation in the Herring Gull (Parsons 1970, 1971) and chicks from large eggs could survive longer without a meal from their parents.

Parent kittiwakes do not automatically feed newly hatched chicks. The onus is on the chicks to employ their innate food-begging technique and beg from parents in the right way and at the right time, so how long chicks can rely upon yolk reserves stored in their bodies could be important for their survival.

These results present some interesting questions regarding geographical variation in size of adult kittiwakes in more northern areas, because these larger birds also lay larger eggs. For example, a large kittiwake egg laid in Britain would be only of average size in the Arctic or in the Pacific. But is the composition of these eggs identical? I suspect that the average yolk contents of these same-sized eggs, laid in the two localities, are different, and that they produce a different sized chick at hatching. This question of size at hatching and at fledging is returned to again when considering the growth of chicks from different areas.

Eggs laid by individual females in successive years

Individual females showed a very strong tendency to lay similar sized eggs in successive years, but age also produced a difference. This pattern is illustrated for three females in Figure 5.6, where all three laid smaller eggs in the first two to four breeding attempts and thereafter produced larger eggs, but of similar size each year. Female 2044892 was below average in wing length and weight for a female, but consistently laid the largest eggs each year in the colony. Perhaps it was not a coincidence that she was also the most successful female in the colony, and she and her partners hatched all of the eggs laid, while only one of their chicks failed to fledge. She also tended to lay slightly smaller eggs in her last four years, but this pattern was not repeated in other females which bred on more than ten occasions, and all three females did not

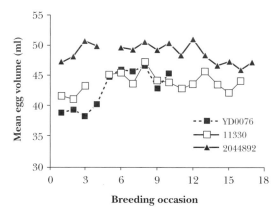

FIG. 5.6   *The average volume of each of the first two eggs laid in a clutch in relation to breeding experience for three individual female kittiwakes. All were two-egg clutches except for occasions 7 to 9 for female YD0076 and occasions 8 to 11 for female 11330, where the mean of the first two eggs of the three-egg clutches has been used. Apart from the trend for egg size to be smaller in the first two to four breeding attempts, individual females tended to lay similar sized eggs in each year. During the period of these records, the mean egg size within the colony as a whole did not change appreciably or significantly and the variation cannot be attributed to year-to-year changes.*

show obvious changes in the size of eggs when they paired with a different partner.

In addition to the pattern shown here, there was a slight tendency for individual females to lay smaller eggs when the date of laying in a particular year was later than usual for the individual, but this effect was typically a difference of less than one millilitre per egg per ten-day delay, and was only of marginal importance.

The shape index (width x 100/length) of eggs laid by individual females also showed a consistent pattern, with some individuals laying eggs which tended to be rounder (higher shape index) than those from other females. As with egg volume, young females tended to lay eggs in which the shape index was low (eggs were relatively long and narrow), and in those females which laid three-egg clutches, the third egg was almost always longer and narrower than either of the first two eggs (Coulson 1963c).

INTERVAL BETWEEN EGGS

In clutches of more than one egg, laying occurs at about two-day intervals but sometimes there may be a delay to a third day. Egg laying can occur at any time of the day, but most are laid during mid-morning and very few are laid overnight or in the evening. The eggs are protected from the time they are laid by being covered by one of the adults, and incubation usually starts about a day after the first egg has been laid, which often results in an asynchronous hatching of the eggs by about a day and occasionally longer. The gap in the time of hatching between the second and third egg is usually about two days, and, on average, is nearly twice that between hatching of the first and second eggs.

FOUR-EGG CLUTCHES

In two of the 36 years of study at North Shields, a four-egg clutch was produced. These were exceptional and were the result of two female–female pairs in which no male was involved. The eggs were incubated but they did not hatch and appeared not to have been fertilised. These two unusual events coincided with a shortage of adult males in the colony, and at the same time one male took two female mates, with each female using different nests within the colony and laying on very similar dates. The male spent two-thirds of the day incubating at one site or the other but, not surprisingly, both females had to do more than half of the incubation. Young were fledged at both sites. Female–female pairs have been reported in the Western Gull (Hunt & Hunt 1977), California Gull and Ring-billed Gull (Conover *et al.* 1979) and in several tern species.

INCUBATION

The incubation period of the kittiwake averages about 27 days from the time that the egg is laid, but it can vary up to two days in individual cases. Much of the variation is caused by differences in the time between the shell first cracking and the chick breaking totally free, which can take more than a day, and also by variation in when intensive incubation first starts after each egg is laid.

Kittiwakes of both species and sexes have three brood patches, despite the infrequent need to incubate three eggs, and both sexes share incubation. Incubation is intensive, probably more so than in most birds. It involves the eggs being covered, protected and warmed by the parents for 99.5% of the time from the laying of the first egg until the last egg is hatched. Apart from the incubating bird standing briefly to turn the eggs, and a brief gap of a few seconds when the pair change over incubating duties, the eggs are covered by an adult.

Maunder and Threlfall (1972) found that the initial development of the embryo in the first egg was slow and Wooller (1973) found that the temperature of the eggs, although continuously covered by an adult, did not climb above 30° C until the day after the last egg of the clutch was laid. As a result, while the development of the embryo in the first egg of a clutch starts as soon as it is laid, the lower temperature and slower initial development rate results in an interval of less than two days between the time of hatching of the first and second eggs, despite the eggs being laid two or even three days apart. Occasionally both chicks in a two-egg clutch hatch on the same day, presumably because of a delay in the start of intensive incubation. In three-egg clutches, the incubating temperature for the first two eggs had been high for about two days prior to the third egg being laid, and as a result the third egg usually hatched two days or more after the second egg, producing a marked hatching asynchrony within the clutch. The reason for the lower temperature at the start of incubation may be due to the vascularisation of the brood patches not being complete at the time the first egg is laid, resulting in a lower rate of heat transfer between the incubating bird and the first egg during the first one or two days. In some cases the first egg is covered by an adult but does not appear to be pressed firmly into a brood patch until the second egg is laid.

Despite the intensive protection and coverage of the eggs by the parent birds, the annual average hatching success at North Shields was lower than the corresponding survival rate of the chicks from hatching until fledging in all but one year of the study (see Figure 9.7). The causes of eggs failing to hatch are several. In the great majority of cases, continuous incubation ceased before hatching could occur, resulting in the death of the embryos. In less than 1% of the clutches, both eggs in the clutch were addled, and about 7% of all eggs laid were apparently infertile. Predation on the eggs was non-existent in most years at North Shields and did not contribute appreciably to the failure rate during the egg stage, which averaged 30%.

Using direct observations to investigate the incubation pattern in kittiwakes and to determine if this contributed to the failure of some pairs to hatch eggs was not a rewarding activity. The incubating bird sits inactive, often asleep for hours on end. Even a four-hour period of observation covered only a sixth of the time spent incubating each day and often it did not involve a single change-over. Further, studying incubating birds at night when it is dark is difficult.

It soon became obvious that we needed an automated method of continuously recording the incubation by each member of the pair throughout the incubation period. I owe much to the late Geoffrey Banbury, an expert in the use of radioactive materials in biological studies at Durham University, for a great deal of help in successfully developing an effective method. The method depended on the fact that a kittiwake at the nest sat or stood in a very confined space, and the leg ring was in the same position throughout each incubation stint. Both members of the pair had a minute amount of radioactive cobalt sealed into the metal leg-ring, with the male having twice as much as the female. The amount of radioactivity was very small, and the radiation produced was comparable to that on the luminous hands of my wrist watch at that time. The radiation from the leg-ring was measured by two Geiger counter tubes placed about 300mm above the nest so that the distance from leg-ring to counter was essentially constant, and so the amount of radiation received differed when the male, female or no birds were at the nest site. The output was recorded as an ink trace onto a chart moving at 50mm per hour. Because of the low level of radiation used, the amounts received fluctuated, but were automatically averaged over two minutes. The setup was as in Figure 5.7. The chart recorded four levels of radiation:

i) a low level of natural background radiation when neither adult was at the nest site,
ii) level x when only the female was present,
iii) level $2x$ when only the male was present,
iv) level $3x$ when the pair were present (sum of the two sources).

This method worked well from the moment of the first trial. The adults were first marked about a week before egg laying and the system operated throughout the incubation period and during the time young were in the nest. Because of the moderately long half-life of radioactive cobalt, some birds were followed for three years, and in the second and third years, they were recorded from their first return to the colony each year and until their final departure from the colony in the autumn. After three years the radioactive cobalt rings were removed. No problems were encountered in using the method and several of the birds continued to breed for several more years without reduced breeding success. In all, seven pairs were similarly marked and their presence at the nest during incubation was followed in detail during 18 breeding attempts.

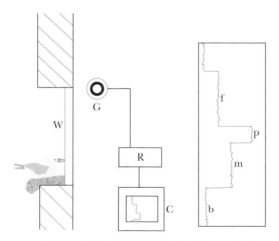

FIG. 5.7 *The system used to monitor the presence of male and female kittiwakes at a nest site using a small amount of radioactive material on the leg rings of the adults, with the male having twice the amount carried by the female. C is the chart recorder, G a Geiger tube in an aluminium casing, R is a ratemeter and W is the window. At the right-hand side is a sample of the chart, with time running from top to bottom and the amount of radiation received by the Geiger tube increasing from left to right. The four different levels represent* b = *background radiation,* f = *female present,* m = *male present,* p = *pair present.*

In passing, it is amusing to mention that a 'problem' of interpretation arose when brief records of levels of radiation higher than $3x$ were recorded prior to egg laying. Eventually, direct observations identified the anomalies as being occasions when the pair mated. In doing so, the male mounted the back of the female and this reduced the distance between the detector and the male's radioactive source, resulting in more radiation reaching the Geiger tube than when the pair was standing side by side. As a result, information on the frequency of copulation was most unexpectedly obtained!

Periodically marking the chart with the time confirmed the times of day for events, so the diurnal pattern of behaviour of the pair during incubation (and at other times) was accurately investigated, with the time of each change-over identified to within a minute. The roles in incubation are described below based on the use of this technique.

Prior to egg laying, the presence of the male and female at the nest changed frequently during the day, as the male in particular made frequent departures and returns associated with courtship, feeding the female and mating. Both members of the pair were present at the nest site for between 15% and 22% of the day and this proportion was appreciably higher than during incubation when they rarely spent more than a total of two minutes per day together as a pair. In general, the sexes shared incubation to a similar extent (overall, females incubated for 54% and males for 46% of the time). Either sex incubated overnight, but females tended to spend slightly more overnight periods incubating. In some bird species where the sexes share incubation, it has been found that only one sex incubates overnight, although this was claimed to be the situation in the kittiwake by Bernhard Hantzsch (quoted in Witherby *et al.* 1943) and is clearly not so.

Immediately following egg laying, time periods spent incubating by each individual ('stints') were relatively short and often of irregular length, but the bird incubating invariably remained on the eggs until its mate arrived, so the timing of the arrival of the relieving bird was the cause of the variability. After a few days, incubation stints become longer but of more similar duration. In the example illustrated in Figure 5.10 on page 94, two days after the completion of the clutch (27 May), eight incubation stints were completed during 24-hours. By 29 May, 2 June and 13 June, these had been reduced to three completed stints each day. During all of this time, each change-over between the members of the pair was rapid, requiring less than a minute. With the exception of two brief visits by the male on 29 May and one on 13 June, only one member of the pair was at the nest site for virtually the whole time, and the pair were together at the nest for less than two minutes of each 24 hour period, i.e. only 0.14% of each day. Figure 5.8 shows the average number of change-overs (and therefore incubation stints) each day in relation to the number of days from the start of incubation averaged for the incubation pattern of 15 pairs. It is evident that the average number of change-overs each day declined markedly during the first five days after the first egg was laid, then continued to decrease slowly for the next ten days and finally remained

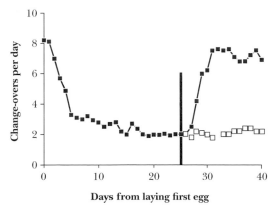

Days from laying first egg

FIG. 5.8 *The average number of change-overs (equal to number of incubation stints) per day during normal incubation by 18 pairs of kittiwakes and also in the period following the hatching (filled squares). The vertical line indicates the usual hatching time. At the start of incubation the change-overs were relatively frequent, but they progressively decreased and soon produced long stints of incubation. After hatching, the frequency of change-overs between the parents increased rapidly, but in three pairs that had addled eggs, the long incubation stints resulted in few daily change-overs (open squares) and this continued for at least a further 14 days after the eggs should have hatched. Based mainly on data analysed by Coulson & Wooller (1984).*

at an average of about two change-overs a day until the eggs hatch. On hatching, the change-overs increase rapidly back towards the level recorded during egg laying, but in three cases where the eggs were addled, incubation continued at the same low level of two change-overs per day for a further 14 days after the eggs might have been expected to have hatched, and at that time the eggs were removed for examination.

The average lengths of incubation stints during the main period of incubation varied between pairs. Most developed a system with an average of two shifts (one each) per day, while others fluctuated between two and three shifts per day, and in one case the pair consistently had four shifts. Usually, the same individual incubated overnight for a run of three or more consecutive nights, but eventually the time of change-over varied sufficiently to result in a change of the individual incubating overnight.

The lengths of incubation stints in individual pairs persisted in subsequent years and continued to differ between pairs, and there was no single or uniform pattern of incubation in kittiwakes, other than that the incubation was shared by both members of the pair. It was obvious that a given pattern was a characteristic of an individual pair, with each pair developing their own pattern of incubation. This was confirmed in two cases where one member of the pair was changed between successive years, when the pattern of incubation became modified in the individual that had been involved in both years. How the pattern of incubation within the pair becomes established remains unknown, but it seems to begin developing in the few days leading up to egg laying.

The time of day when change-overs occurred is shown in Figure 5.9A, and the pattern is clearly bimodal. The starting times of incubation stints which continued entirely in daylight were unimodal, and these formed the well-defined peak between 05.00 and 07.00 hrs. On the other hand, change-overs which involved the incoming bird incubating overnight took place after midday and these stints lasted about twice as long as those where the change-over took place before 11.00 hrs in the morning (Table 5.4), with many absences continuing for up to 16 hours.

If an incubation stint lasted overnight, the off-duty bird remained at sea and did not return to the nest at first light (about 03.30 hrs) but delayed arriving until between 05.00 and 07.00 hrs. Why was there this delay? It seems that the off-duty individuals did not 'count', and perhaps did not actively use the period between sunset and sunrise (about six hours in late May at North Shields) as part of the off-duty period, perhaps roosting and sleeping well out to sea, then feeding at and after dawn and then finally flying from the off-shore area for one or two hours to return to the colony.

The times of change-overs when chicks were in the nest (Figure 5.9B) differed from those during incubation (Figure 5.9A). The pattern of change-overs during the chick stage did not show a marked bimodality, probably because the duration of stints at the nest, and so the time away feeding, were shorter. By this point in the breeding season, the duration of daylight was also longer, but reduced activity during the hours of darkness was still evident.

The lengths of overnight incubation stints were an average of two hours longer when females rather than males incubated (Table 5.4). This means that when females were off-duty overnight they tended to return to the colony more than an hour earlier in the morning. Perhaps when off-duty overnight, the females did not travel as far to feed as did the males and so could return earlier to relieve their partner? With the introduction of loggers which record the movement of birds, these questions may be soon answered.

During incubation, kittiwake pairs spend remarkably little time together at the nest; both sexes spend about 12 hours each day away from the colony. Despite the pair spending much time together during the pre-laying period, this length of time away from the colony is similar to that during incubation, but involves a different pattern of behaviour. Prior to laying, both members of the pair leave the colony in the evening, return during the morning, and spend several hours together as a pair at the nest site during the day. So the low proportion of pairs present in the colony during incubation does not indicate that food is any more difficult to obtain at this time. The birds are simply using different patterns of behaviour.

As a result of this study, three fundamental 'rules' concerning incubation became evident. First, the incubating bird remained on the eggs until its partner returned, and the eggs were never deserted. The second rule was that if a relieving bird had not returned by 22.00 hrs, the arrival was delayed by a further 6–8 hours, and until between 05.00 and 07.00 hrs of the next day. The third was that individual pairs organised their own patterns and durations of

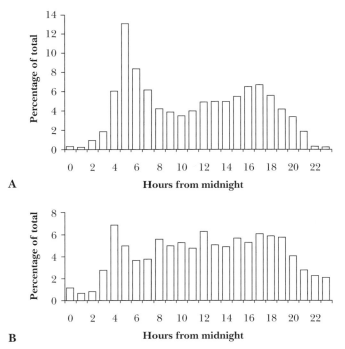

FIG. 5.9 *The time of day that adult kittiwakes changed over nest duties.* **A.** *The times of change-over during incubation. Note the marked peak between 05.00 hrs and 07.00 hrs when most individuals incubating overnight were relieved.* **B.** *The times of change over during chick rearing (when the length of the night was shorter than during incubation). In both periods there were relatively few change-overs in the late evening and before sunrise, but the diurnal pattern was more appreciably pronounced during incubation. The percentages shown are for change-overs in the 60-minute periods starting at the hour shown. Based on an analysis made by R. D. Wooller.*

TABLE 5.4 *The mean duration of incubation stints by male and female kittiwakes, according to time of start. The morning stints started between 00.00 hrs and 11.00 hrs. Afternoon stints started after 12.00 hrs and were completed before midnight. Evening stints involved the same bird continuing incubation overnight.*

Start period	Male		Female		Significance
	N	*Mean (hours)*	*N*	*Mean (hours)*	
Evening	113	13½	154	15½	P<0.01
Morning	144	7½	113	6½	P<0.05
Afternoon	73	4½	62	4½	n.s.

incubation shifts, and typically these resulted in very similar effort by both members.

A case in which eggs failed to hatch is shown in Figure 5.11. The presence of the sexes at the nest during incubation is shown from the time the clutch was completed. For the first seven days of incubation, the shift system operated as expected, with progressively longer stints gradually developing into a system averaging two change-overs per day. However, eight days after the clutch was completed, the male failed to return at the expected time to carry out his next stint of incubation, but the female compensated for this and incubated continuously for a day and a half. On day 9, the male returned (a day late!) and the change-over and departure of the female took place within a minute of the male's arrival, as normal. The female returned after seven hours and again took on the overnight stint, but the male did not return during the next two days. The female compensated for this and incubated

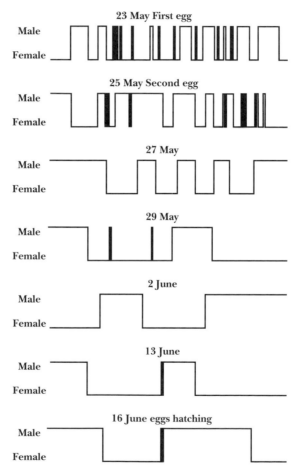

FIG. 5.10   *The presence of male and female kittiwakes at the nest during incubation. The illustration runs horizontally for 24 hours from midnight to midnight on each date. The first egg was laid on 23 May and the second on 25 May, during which time there was frequent arrival and departure of the members of the pair. By 27 May, an incubation pattern had been developed, with eight stints of incubation during the 24 hours, but by 29 May to 13 June, the incubation stints were reduced to four stints per day. Note that either sex incubates overnight. The nest site was not left unoccupied.*

continuously for all of this time, and until the male reappeared on day 12. He quickly took over incubation and the female left immediately only to return and to take over incubation after an hour. Within a minute of the female arriving the male left again and did not return for three days. The female incubated continuously for 48 hours, briefly left the nest and eggs for about five minutes, and then incubated continuously for a further day. At this point, the male returned, did a normal daylight stint of incubation and then disappeared for a week. For the first three days of the male's absence, the female incubated continuously, but thereafter left the nest and eggs on several occasions. Nevertheless, she was present at the nest for over 93% of the time the male was away and clearly she endeavoured to compensate for his absence. The eggs failed to hatch and upon examining these later, the embryos had died when about two weeks old and probably during the period of abnormal behaviour by the male.

This same pair nested together in the following year. Incubation was normal for the first 22 days, with two change-overs each day from day 10 to day 22. But the pattern was already abnormal because the female incubated overnight in all twelve 24-hour periods, thus doing the greater part of the incubation. The male did not arrive at the nest site at all on days 23 to 25, and the female compensated by incubating continuously for 60 hours. The male appeared and relieved the female briefly on day 26 but was not recorded there again during the next 16 days, while the female occupied the site, both

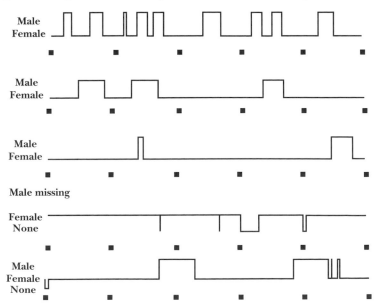

FIG. 5.11   *The roles during incubation of the kittiwake male and female on site WIG at North Shields, illustrating the aberrant behaviour of the male and the considerable effort made by the female to compensate for his absence. Each square denotes midnight and each line covers five days.*

by day and by night, for 85% of this time and thereafter visited the site less frequently. The female had ceased to incubate by day 30, and the male was never seen again. The pair was not inexperienced, and the male had reared ten chicks in eight previous breeding seasons and also had bred successfully in an earlier year with the female involved in this study. It is only possible to speculate that during this study he encountered health problems which prevented him from being able to integrate into a successful incubation pattern, and he died soon after.

It should be noted that while the male's behaviour in these two cases was atypical of breeding kittiwakes, the female in one of these cases also showed unusual behaviour in that she returned to the nest only an hour after having been relieved by the male following an exceptionally long incubation stint which lasted two days. Why did she return so soon?

It is evident that the incubation system is variable in the kittiwake and the exact pattern is decided by each pair. How this is developed is not known, but it can be a source of difficulty and may result in a failure to incubate satisfactorily in some cases. What became apparent was that individuals can and did compensate to a remarkable extent for the inadequacy of their partner.

CHAPTER 6

Chicks, growth, fledging, productivity and desertion of the colony

THE rearing of kittiwake chicks involves the parents changing from a pattern of incubation behaviour to one suitable for brooding, protecting and bringing food to and feeding the chicks. Initially, hatching does not cause major changes in the behaviour of the parents, since instead of incubating the eggs, one parent remains at the nest to brood the chicks while the off-duty individual still spends the time away from the colony. The presence of both parents at the nest is still infrequent.

At North Shields, kittiwake chicks weighed an average of 33g at hatching but this varied with the size of the egg from which they hatched. Irrespective of the size of the egg, each chick weighed 15g less than the weight of the egg when it was laid. Only about 2.9g of this loss can be attributed to the weight of the shell and the rest is due to loss of water from the egg during incubation and the consumption of yolk for respiration and the development of the embryonic chick. As a result of these losses, an egg weighing 50g (46.3ml volume) tended to produce a chick of 35g, while a small egg weighing 40g (37ml) hatched a chick weighing 28g or about 20% lighter. However, a 'small' chick was not structurally as small as might be expected from these weight differences, and examination of the semi-transparent abdomen suggested that the main difference was due to greater amounts of yolk being retained within the body of chicks hatching from larger eggs. This additional yolk in chicks hatched from larger eggs almost certainly has a major survival value, allowing the chick more time in which to establish the all-important ability to beg for food and persuade its parents to feed it. Without active begging by the chick, parents will not feed it.

When first hatched, a kittiwake chick is covered in light grey down and has its eyes open. It is unable to regulate its own body temperature, and heat must be gained from the parents, as in incubation, to keep it alive and active. For the first few days after hatching, the long shift system used by the parents during incubation continues and the chicks are brooded intensively. At this time the chick needs and receives relatively little food, and so this is not a great demand on the parents, but keeping the chick's body temperature high is essential to preventing it becoming moribund. During its first week of life and as it grows, the chick progressively develops its own ability to maintain and regulate its body temperature and requires less brooding. After the first week out of the egg, the chick is rarely brooded unless the weather is cold, but nevertheless a parent remains continuously at the nest site to protect and feed the chick.

Other changes in the behaviour of the parents develop during the chick's first week. As shown in Figure 6.1, the duration of stints guarding and brooding two chicks at the nest became progressively shorter, and by the time the older chick was two days old, the number of change-overs by the parents during 24 hours had doubled, although a long overnight stint at the nest, coupled with an equally long absence by the off-duty parent, still occurred. Thereafter, overnight stints became shorter, and this is also evident in Figure 5.2, where the timing of change-overs no longer shows the marked peak of off-duty birds returning to the nests between 05.00 hrs and 07.00 hrs which is so characteristic of the incubation pattern. Nevertheless, parents still infrequently returned to the nest during the (shorter) periods of darkness. By the time the older chick was 13 days old (30 June), the overnight attendance by one of the parents at the nest was shorter and most of the daytime stints were briefer, although on one occasion the male remained away for six hours. This change to several shorter stints at the nest and one longer absence each day, usually by the male, was typical of chick-rearing and is well illustrated by the shift system on the day that the older chick was 33 days old and both chicks were approaching fledging. Presumably, these changes in behaviour are associated with the need to obtain food and feed the chicks several times a day. It is not clear why the long absences by the parents, which occurred frequently during incubation, became much reduced. Do the parents feed nearer the

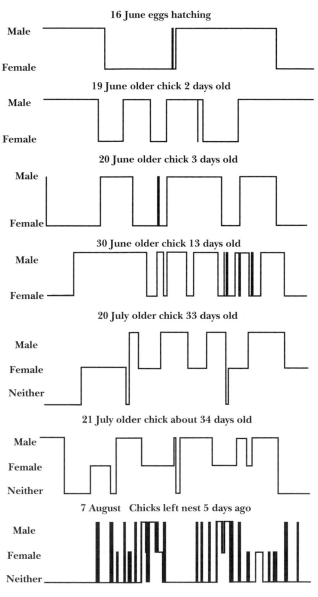

FIG. 6.1   *The pattern of stints of attendance of male and female kittiwakes at hatching of their chicks, during the nestling stage, and (in the last case), five days after chicks had left the nest. The width of the figure represents a 24-hour period from midnight to midnight. Three different levels of horizontal lines from top down indicate the presence of the male, the female and neither partner was at the nest site. Both members of the pair were together at the nest site for such brief times (usually less than a minute) during each change-over that they are not shown on the scale used. The first occasions on which chicks were left unguarded were on 20 July when they were about 33 days old. The first occurred overnight and was caused by the female leaving the chicks, as was the second, while the third was caused by the male departing.*

colony? Are smaller fish now required and have these become more abundant near the colony as a result of early spring spawning near the coast, producing young fish (0-year class sandeels) suitable as kittiwake food? Samples of food from kittiwakes indicate that the adults switch from capturing larger to smaller sandeels at about the time chicks hatch. As a result, a successful spring-breeding season by fish, particularly sandeels and Capelin, may be important in the successful breeding strategy of kittiwakes.

WHEN ARE THE CHICKS LEFT ALONE AT THE NEST?

The continuous recordings of radioactive-tagged pairs ensured that the first time chicks were left alone at the nest was accurately known. This method does not have the errors, which can be appreciable, arising from spot observations made throughout the day and which can overestimate the age of chicks when they are first left alone at the nest (Coulson & Johnson 1993). The magnitude of error arising from spot observations is illustrated in Table 6.1, which gives the average delay in identifying the first time chicks were left unattended using different numbers of random records taken each day and then comparing those with the date of actual first occasion detected by the continuous records.

The continuous records of the presence of parents at the nest produced reliable data on the time of day that the chicks were first left alone, and this also included the night (dark) period and very early hours of the morning. A few spot observations each day resulted in greatly overestimating the age at which the chicks are first left unattended. Even with 24 spot observations per day, the first recording of the actual date when the chicks were first left unattended is still almost three days after the first event actually occurred. In addition, 85% of the occasions when the chicks were left unattended for the first time were during darkness, and so these would be missed by visual observations restricted to daylight hours. In some instances, the first daylight absence of both parents from the nest was more than seven days after their first nocturnal absence. It is evident that the use of spot observations greatly underestimates the time when chicks are first left unattended by adults.

In the pair recorded in Figure 6.1, one parent was always present at the nest until the older chick was 33 days old. In a second case, chicks were never left

TABLE 6.1 *The average delay in days which is likely to occur in recording the first occasion on which broods of kittiwakes were left unattended if based on different numbers of daily spot observations, and when the first absence of both parents was detected by the continuous records from the radio-tagged parents.*

Spot observations /day	1	2	4	12	24
Delay (days)	10.5	9.3	7.7	6.1	2.8
Delay (days) if spot observations were restricted from 06.00 hrs to 22.00 hrs	13.3	11.4	8.7	7.7	4.2

unattended for 40 days, that is, about four days after the chicks would have been capable of flying. The continuous presence of one or other of the adults at these nests extended from egg-laying for 59 and 67 consecutive days. In other pairs, an adult remained continuously with the chicks for shorter periods of time, and Figure 6.2 shows a range of examples of when adult kittiwakes left their chicks alone for the first time. The first absences are of short duration, often only a few minutes, and are caused by the bird in attendance breaking the 'rule' which operated from the start of incubation, namely that one bird remains at the nest until the return of its partner.

Figure 6.2 shows that there is appreciable similarity between the time of first desertion of the chicks by each of the parents. When one parent leaves the chick unattended, then the other tends to do so within a day or so, and it is interesting to speculate how this synchrony occurs. In only some cases did a returning parent find its partner absent from the nest. In others, the individual leaving the chicks alone returned to the nest before its mate arrived, so presumably, the mate was not always aware that its partner had changed its behaviour and temporarily deserted the nest and chicks. So in these cases, how did the other parent know the chicks had been left alone?

One possibility is that both parents were reacting to the behaviour of the chicks, for example, by the intensity of their demands for food. Do hungry chicks compel parents to leave them unguarded at an earlier age? If so, it is reasonable to suggest that the age at which young kittiwakes are first left alone on the nest is a measure of the ease or difficulty the parents are having in finding food. If food is scarce (or the parents are poor hunters), both parents would need to search for food at the same time if the chicks were to survive, and the complementary system of attendance at the nest would have to cease. The presence of unattended young at an early age could indicate food shortage, and counts have been used of the proportion of nests with chicks but without parents as an index of this effect. In general, there seems to be a relationship between how early chicks are left unattended and the breeding success in a colony, but a few spot estimates of the proportion of nests without parents present are unlikely to be an accurate measure of the extent of the temporary desertion of the chicks, although such data might be of value in year-to-year comparisons.

By comparing the behaviour of three pairs in the same year (Figure 6.3), it is evident that the time of first absence (and then the more frequent absence in relation to chick age) varied considerably from pair to pair nesting in the same colony and virtually at the same time. Figure 6.3 shows the proportion of each day during which a parent was present at each of three nests from the time of hatching until the chicks finally left the nest. In these three examples, the chicks were continuously guarded by one or other of the parents for 29, 20 and 14 days. In Figure 6.3A, the first absence was followed by a further three days without another absence, and then a progressive decline in attendance until the chicks were guarded for only 30% of the time. In the second example (B), one of the parents guarded the chicks for 20 days and then for

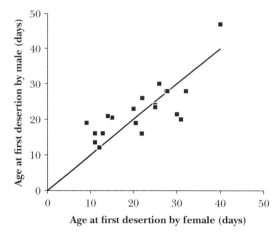

FIG. 6.2   *The age of the older chick when male and female kittiwake partners temporarily deserted the chicks and left them unattended for the first time. The parent that was at the nest and then left before his/her partner had returned was regarded as the individual which deserted the chicks. The line shows the trend if both partners behaved similarly and both left the chicks unguarded on the same day. Note the variation between individual pairs.*

over 90% of the time for a further nine days before the main decline in attendance occurred, eventually reaching a presence of less than 25% of the time. In the third example (C), total attendance of a guarding adult continued for 14 days, remained at between 80% and 90% for a further nine days after the first absence before the main decline took place, and reached less than 20% attendance for 14 of the last 15 days for which the chicks were at the nest. Neither the timing nor the extent of the incomplete guarding influenced the age at which the chicks fledged.

So why were the pairs different in chick-guarding in the same year and at virtually the same time? Food availability would not readily explain the difference since the effects were very different in all three pairs and at similar stages of breeding. The most likely explanation is that the drive to protect the chicks differed in the three pairs, but then why did both members of the pair behave similarly? Were those parents that persisted longest in protection better-quality individuals that needed less time to obtain food for themselves and their chicks? Or did some parents find the pestering for food by the chicks caused them to delay returning to the nest, although they remained in the vicinity?

The behaviour of the parents also begs the question as to the purpose of their presence. In the colony studied, chicks were not at risk from predation, but this could be important in some other kittiwake colonies. It is possible that the presence of a parent had nothing to do with the risk of the chicks being taken by predators. Perhaps their presence prevented prospecting sub-adults from landing and taking over possession of the site, something that is attempted in all kittiwake colonies. Prospecting adults do not attack chicks

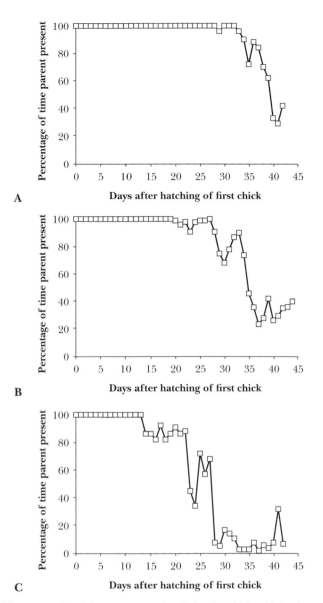

A

B

C

FIG. 6.3   *Three examples of the percentage of each day for which a kittiwake parent was present with the chicks at three nest sites, shown in relation to the age of the chicks and until the last chick left the nest permanently. A hatched on 8 June , B on 10 June and C on 4 June, all in 1970.*

in the absence of the parents, but if they managed to land repeatedly on the site they could gain in confidence, take possession of it more frequently and challenge the parents' site ownership in the following year. It may be of significance to note that sub-adults do not visit the colony at night, the time at which most chicks are first deserted by their parents.

THE DURATION OF FEEDING TRIPS
DURING CHICK REARING

During most of the daylight hours, absences from the colony by parents with chicks averaged about 2 hours 48 minutes (n = 4,348, s.e. 1.6 minutes). With a flight speed of 45 km/hour, this gives them an average feeding range from the colony of under 63km (40 miles). This value is nearly identical with the 2 hours 38 minutes, based on 158 feeding trips, recorded on the Farne Islands by Pearson (1968), and as a result, he estimated a very similar feeding range. At North Shields, departures in the late evening involved the individual remaining away during the hours of darkness, and these trips lasted about twice as long and so gave the individual a potential feeding range double that used during most of the day (Figure 6.4). I suspect that the first part of the dark hours are spent inactive on the sea and that these birds did not normally make appreciably longer journeys from the colony. There is still considerable uncertainty as to the extent to which kittiwakes feed at night. The use of loggers which record the position and behaviour of individuals will soon rectify this lack of data and give additional information as to for how much of the night kittiwakes search for food and how far from the colony they travel.

With a coastal colony, where birds are able to feed within 180° from the colony, the kittiwakes have a maximum of about 20,000sq km (7,700sq miles) of sea available to obtain food each day. In a colony of 5,000 pairs, this represents a kittiwake for every 2sq km (0.77sq miles) of sea if all were feeding. This is a low density and is unlikely to have an appreciable impact on the fish stocks in the area. The median kittiwake colony size is only about 300 pairs. Any appreciable impact on fish stocks in areas exploited by kittiwakes might only occur in the few largest colonies, say those with over 25,000 pairs, or where there are several medium-sized colonies close together, but it is unlikely to apply in the typical colony.

FEEDS PER DAY

Pearson (1968) measured feeds per day given to individual chicks on the Farne Islands. Feeds averaged 4.6 per day in chicks up to 20 days old (and involved amounts increasing to about 125g of fish per day). This level of feeding dropped to about 3.5 feeds a day in older chicks and to 86g of fish each day as chicks approached fledging. These figures produce a rough estimate of each chick receiving about 3,000g of fish until it is capable of flight at 36 days old. During this time, the chick grows from about 33g at hatching to 350g. With a food absorbance efficiency of 80%, the growth of the chick requires 540g of this, while the remaining 2,400g (approximately 1,900g absorbed) is used to maintain body functions including regulation of body temperature and the costs of digestion. Another estimate of the total food required by chicks up to 36 days, based on data in Coulson & Porter (1985), was 3,200g. The precise amount is modified by variation of the calorific value of the particular food consumed.

In both cases, about three-quarters of the food consumed was used for maintaining body functions and only a minority of it then went into the actual growth of the chick, with the former taking priority over growth. If anyone has any doubt about the high proportion allocated to maintenance, I would draw attention to the large amount of food I fed to orphaned kittiwake chicks each day before they returned to the maximum weight they had reached on the previous day, and before they put on additional weight. It took four feeds per day for the chicks to regain the weight of the previous day, and only when they were given a fifth feed did they increase in weight.

Maintaining normal body functions, including temperature and digestion, is the first priority for the use of assimilated food, and only after the food consumed and digested meets these needs is growth achieved. As a result, measuring the growth of a chick is an extremely sensitive way of measuring the amount of food it receives. A decrease of 10% in the average daily growth rate of a chick represents only about 2g less food fed to the chick each day, or approximately a 2% decrease in the average daily intake. If the food intake of the chick was reduced by 11%, this is likely to reduce the daily growth rate of the chick by 50%, for example from a gain of about 16g/day to 8g/day.

FLEDGING PERIOD

Kittiwake chicks nearing fledging do not make the leaps into the air accompanied by intense wing flapping so commonly seen in ground-nesting gull chicks as they near fledging. There is too much risk of the chick being

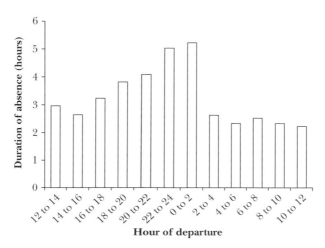

FIG. 6.4   *The average duration of the absence of kittiwake parents on feeding trips while rearing chicks in relation to the time of departure from the nest site. Based on 1,542 trips.*

displaced by wind and updrafts on the cliff face, thus preventing it from remaining on the nest site. Kittiwake chicks exercise their wings by facing inward and holding firmly onto the nest with their curved claws. They only occasionally lift off the nest, and then for only a few centimetres and for a few moments. Kittiwake chicks are at least 36 days old before they can sustain flight. In most cases, chicks leave the nest sometime after day 36 and then return again within a brief period. I have not seen chicks attempting to feed themselves during these brief excursions. The chicks spend the time away from the nest flying around the colony and making attempts, which are often futile, to land on suitable cliff sites. However, they are only fed by their parents at their nest, so if they land elsewhere in the colony, they do not receive food. Records of the first observed absence of a chick from a nest (fledging period) averaged 41.5 days after hatching (Table 6.2). This period is comparable with those reported earlier by Coulson & White (1958a) and by Maunder & Threlfall (1972) of 42.7 days and 41.6 days, respectively. No differences were found between the fledging periods of single-chick and two-chick broods.

On average, chicks finally leave the colony ten days after the first flight, but in extreme cases, fledged chicks may return to the nest for up to 61 days after hatching. The chicks are not fed by their parents away from the nest site, unlike the situation in many other gull species where chicks and parents leave the colony together. In the kittiwake, the final departure of chicks abruptly breaks the bond with their parents and from that time they have to forage for themselves. The young produced by the last few kittiwakes to breed in the colony each year do not usually benefit from this period of returning to the nest and in most cases the first departure is also their last, averaging about 45 days after hatching. Presumably this is a disadvantage to the survival of young produced late in the breeding season.

The final departure involves the chick flying alone directly away from the colony and out to sea. Very few of the independent young remain within sight of the colony, and they disperse rapidly into off-shore waters. Ringing recoveries show that some have moved over 4,000km within six weeks after leaving the colony. Sometimes these young birds temporarily join feeding groups of adult kittiwakes and perhaps they benefit by copying them. This sudden inde-

TABLE 6.2 *The number of days from hatching to fledging (first flight) and to the final departure of 134 kittiwake chicks, based on intensive checks made throughout each day. There were no significant differences between those in broods of one and two chicks. Data from A.F. Hodges (1974) and the author.*

Length (days)	32–34	35–37	38–40	41–43	44–46	47–49
First flight	0	11	39	57	14	12
Final departure	0	0	1	3	11	22

Length (days)	50–52	53–55	56–58	59–61	Total	Mean
First flight	1	0	0	0	134	41.5
Final departure	48	27	19	3	134	51.3

pendence of the young kittiwake must be a critical time, particularly since they do not seem to have accumulated large food reserves before departure, and usually they have already dropped below the asymptotic weight they reached as pre-fledged chicks.

CHICK GROWTH

The pattern of growth in kittiwake chicks is similar to that of most birds. Initially, there is a slow increase in weight, followed by an appreciable period during which time the daily gain in weight is virtually constant. Eventually, the daily gain progressively decreases and an asymptotic weight is reached which is similar to that of the adults. This maximum weight coincides with the

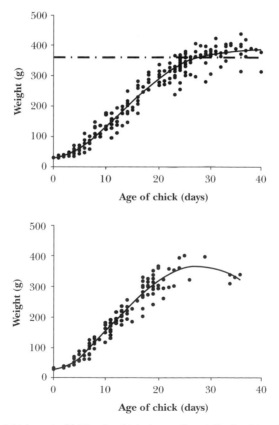

FIG. 6.5   *The weight (mass) of kittiwake chicks in north-east England in relation to their age. The upper graph is from North Shields in 1978, and the lower graph from Gateshead in 1995. A third-order polynomial has been fitted to the points. The horizontal dashed line in the upper graph indicates the average weight of breeding adults at the colony. The mean growth rate between 75g and 300g is 16.7g/day in both cases.*

approach of fledging and the final growth of plumage, particularly the flight feathers. In some individuals, this maximum weight is maintained but in others, there is a modest loss prior to fledging. Generalised growth curves for the kittiwake at North Shields and Gateshead, north-east England, are shown in Figure 6.5. They show a virtually constant daily mass increase between the time the chick weighs 75g and 300g. As a result, it is often convenient to disregard the very early and late growth and record the daily average for the linear part of the weight increase. This linear part of the growth curve can be measured by reducing the handling of chicks to weighing on only two occasions, two weeks apart, say from when the chicks are about 6 days old until they are 20 days old and thus reducing the chance of the chick regurgitating its last meal and possibly reducing growth.

It is typical of the growth pattern of chicks that feeds provided by parents in the first few days after hatching are infrequent, and the chicks continue to depend in part on the stored food remaining in what had been the yolk of the egg. This reserve can be important because it allows the chick and its parents time to develop the interactions of food begging, which results in the chick obtaining food from the throat of the parent. While the first chick to hatch develops this relationship with its parents within a day or so, and is soon successful in both begging and obtaining food, it often takes longer for the second chick to develop the relationship. As the second chick usually hatches a day or so after the first, this delay in feeding sometimes produces a greater difference in the time when the growth of the second chick starts, resulting in a more noticeable difference in size within a brood of two than is attributable to the difference in hatching dates. If there is a third chick, this individual has even more difficulty in establishing a relationship with its parents and obtaining its first meal. It often hatches three days after the first chick, and the start of its growth can delayed for two or three days more. This exaggerates the weight difference between the chicks in the brood (Figure 6.6A and B). However, despite this initial delay, the second chick to hatch, and the third when it occurs, can grow at a similar rate to the oldest sibling (Figure 6.6A). As a result, the later chicks in a brood often have the same growth curve as the first chick to hatch, apart from the time at which the actual growth starts. A similar pattern was been found by Langham (1972) between the brood members of Common Terns.

This delay in starting growth is a particular handicap to the third chick because it hatches from a small egg and, as a result, has a smaller yolk reserve after hatching. The delay in establishing 'recognition' by its parents often results in the early death of the third chick, and this occurs days before the food needs of the brood become appreciable. So the early death of the third chick should not be attributed to food shortage, but rather to the difficulty in being able to beg effectively enough to persuade the parents to feed it despite competition from the vigorous begging of its larger siblings. While the survival of the third chick is lower than for others in the brood, it remains the case, as shown in Chapter 5, that clutches of three eggs result in

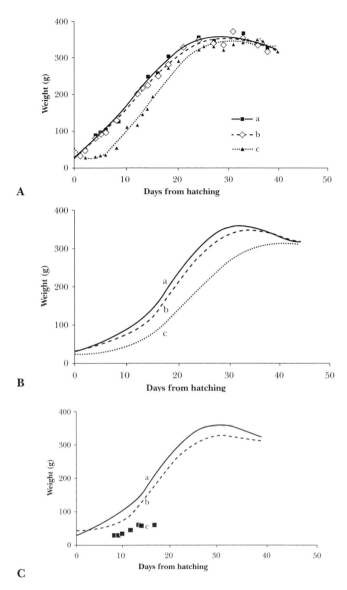

FIG. 6.6 *Typical variation in the growth curves for three different kittiwake broods of three chicks at North Shields from the time that each hatched. It is evident in each case that the start of growth of the third (c) chicks is delayed. In example A, the c-chick eventually obtains more food each day than the first two chicks to hatch and catches up on its growth by having the highest growth rate during the linear part of the growth. However, in example B, the third chick failed to obtain as much food each day as the older chick but the difference was small (perhaps only 3–4g of fish per day) and it reached a fledging mass similar to the other two. In example C, the a- and b-chicks grew at a similar rate, but the c-chick grew very slowly and failed to gain any weight during the last five days before it died on day 13. However, for the first six days it did obtain sufficient food to keep it alive and to put on some weight.*

appreciably more young fledging per clutch than from two-egg or single-egg clutches.

SIBLING RIVALRY

Over the years, I saw only minor evidence of intensive sibling rivalry among broods at North Shields, and this was mainly evident in broods of three chicks. I suspect that this was because food was plentiful throughout the study, and support for this comes from the very high fledging success in every year. If a chick died, it was likely to be in the first few days of its life, but whether this was caused by sibling rivalry or the result of several other possible causes is not clear. When well fed, the first hatched chick was mainly quiescent and did not create a problem for its siblings. In contrast, an underweight chick was active, vocal and aggressive towards the handler, and it would be easy to see how this aggression could be directed towards a sibling. When chicks are brooded, the opportunities for sibling aggression might be expected to be reduced because the presence of the brooding adult greatly limits movement by the chicks and, if it was to occur, it is more likely to be evident when the intensity of brooding is reduced. But observations in colonies where food shortage occurs indicate that this was not so (White *et al.* 2010), and the frequency and intensity of aggression were highest in the first week of life. That this aggression was closely linked to food shortage was also revealed in the study of White *et al.*, with an appreciably lower frequency of aggression occurring when food supplementation was introduced.

Clearly, kittiwakes can and do reduce their brood size when food is insufficient through sibling rivalry, and this often reduces the brood size to a single chick in localities where and when food shortage occurs.

WHAT LIMITS THE GROWTH RATE OF CHICKS?

If the daily growth rates during the period of maximum weight increase of chicks which successfully fledged in a series of gull and tern species are analysed, there is clearly (and not surprisingly) a higher daily growth rate in the species where the adults are larger and heavier. However, if this rate is converted to the daily gain as a percentage of adult weight, the relative growth rate is lower in the larger species (Figure 6.7). One of the consequences of this relationship is that, in general, larger species have to sustain a longer period of growth and this results in a longer fledging period.

The comparison of the eight species in Figure 6.7 shows the logarithmic relationship between the maximum daily increase in weight of chicks as a percentage of the adult mass and the size of the adults, and the kittiwake fits into this relationship. But why do the young of large species take longer to grow? The answer probably lies in the capacity of the stomach and the time

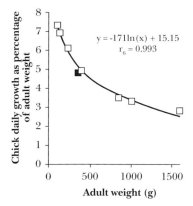

FIG. 6.7   *The peak daily growth of chicks of terns and gulls as a percentage of the adult weight in relation to adult weight. Points from left to right are for Arctic Tern, Common Tern, Sandwich Tern, kittiwake (filled square), Common Gull, Lesser Black-backed Gull, Herring Gull, and Great Black-backed Gull. The points closely fit a logarithmic curve. Chicks of larger species grow slower than those of smaller species.*

taken to digest the food, and the fact that these are not related linearly with adult weight. In the case of the kittiwake, the limit imposed by its size results in the chick being unable to ingest, digest and assimilate in a day any more food than can produce a growth of about 17–18g per day. The existence of this limit is supported by the fact that single chicks do not have a higher growth rate than those in broods of two (Table 6.3), although obviously the parents could feed more food to a single chick.

It is possible that on occasion, growth rates larger than 18g per day may be recorded, but these probably arise from cases in which the chick was weighed twice, with the first weighing taken some time after the last meal and the second weighing made shortly after a meal. In effect, there is a physical limit to the maximum growth rate that chicks can achieve. The existence of this limit is an advantage to the second and third hatched chicks, since once the first chick is satiated, the younger chicks have reduced completion and an easier chance of obtaining food from their parents.

TABLE 6.3   *The daily increase in weight of young kittiwakes during the linear part of the growth in relation to the brood size and the order of hatching. Cases when one (or more) of the chicks in a brood died are not included. ** indicate significant differences from the other members of the brood. The data for the chicks in a brood of three are based only on those where all survived for the first 20 days after hatching.*

	Brood 1	Brood 2a	Brood 2b	Brood 3a	Brood 3b	Brood 3c
Number	232	321	321	40	40	40
Mean g/day	16.5	16.5	15.2**	18.1	16.0**	13.9**
S.E.	0.20	0.14	0.15	0.51	0.58	0.66

In addition to abundant data on the growth of kittiwake chicks from the UK, there is also extensive information on the growth of the Pacific subspecies in colonies in Alaska, and a growth curve based on these data is shown in Figure 6.8. Kittiwakes in Alaska are appreciably larger than those at North Shields, and to reach this 10% higher asymptotic weight in the Pacific it might be expected that the chicks there would show a somewhat higher daily growth rate than in the UK. The eggs of kittiwakes in Alaska are larger than those at North Shields and the chicks hatch with about 11% greater weight. Despite this, and the need for Pacific birds to reach a higher weight, the growth rates in the two areas (in terms of actual weight increase) are not distinguishable during the first 20 days of the chick's life (Figure 6.9). The Pacific kittiwakes reach the higher asymptotic weight only by continuing their growth for a longer number of days. It is possible that the fledging period in the Pacific is longer, but there are too few data to decide whether this is so.

GROWTH OF RED-LEGGED KITTIWAKE CHICKS

There is only one detailed study of the growth of the chicks of the Red-legged Kittiwake and that was made on St George Island in Alaska in 1993 (Lance & Roby 1998). The growth curve is very similar to that of the Black-legged Kittiwake, and Lance & Roby obtained a maximum growth rate of 15.68g/day. However, this was lower than the comparative rate for the Black-legged Kittiwakes in the same colony in the same year, recorded as 18.76g/day. From this, it might be presumed that the Red-legged Kittiwake had a poorer daily weight increase presumably because it had more difficulty in obtaining food, but this was not the case, as the Red-legged Kittiwake adults are smaller than those of the Black-legged Kittiwake in Alaska. However, the adult Black-legged Kittiwakes at North Shields have similar weights to the Red-legged Kittiwake and had a similar daily growth rate. Lance & Roby (2000) also swapped chicks between the species, but the samples were small and did not show any meaningful differences between the growth rates of chicks reared by their own parents and by foster parents.

Further investigations in Lance & Roby's study indicated that the Red-legged Kittiwake chick was fed at only slightly above half the frequency found for the Black-legged Kittiwake chick, despite meals for each species being of a similar size, but meals were composed of different food species. While the food supplied to Red-legged chicks had a slightly higher calorific density of 21.6kJ/g compared to 18.4kJ/g (Lance & Roby 2000), this difference is far too small to compensate for the appreciably fewer meals per day.

There would appear to be an anomaly within these results. Such a large difference in provisioning rates of food between the species, and with only a small difference in the calorific density of the food, should have resulted in no growth whatsoever in the Red-legged Kittiwake chicks and produced acute starvation as all of the food supplied would be needed to meet the body's

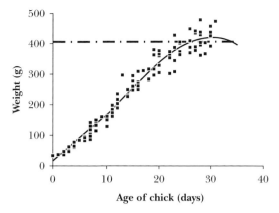

FIG. 6.8   *The weight (mass) of kittiwake chicks in relation to their age in Alaska. Data supplied by D. G. Roseneau. A third-order polynomial has been fitted to the data points and shows a growth rate of 16.9g/day between 75g and 300g. The dashed line is an average weight of an adult breeding in the same colony.*

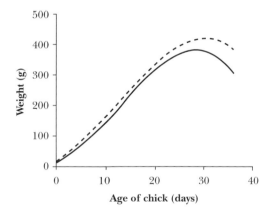

FIG. 6.9   *The average growth curve of kittiwake chicks at North Shields (continuous line) and in Alaska (dashed line, taken from Figure 6.8). The chicks in Alaska were 11% larger at hatching, but until about 20 days old, there was virtually no difference in the daily increase in weight of chicks from the two areas. However, those from Alaska increased in weight for a longer time than those from north-east England to reach the higher asymptotic weight, rather than having a greater daily growth rate between 75g and 300g to reach the asymptotic weight.*

metabolic needs, leaving none for growth. I calculate that the lower daily growth rate of the Red-legged Kittiwake chicks compared with those of the Black-legged Kittiwake represents only some 5g less food being consumed by the chick each day, that is, a reduction of less than 7% in the provisioning. As a result, it seems that one of the measures Lance and Roby obtained is incorrect. They measured provisioning rates from samples of observations made only between 06.00 hrs and 24.00 hrs and I wonder if they failed to consider that the Red-legged Kittiwake, which is more active in the dark than the Black-legged Kittiwake, probably fed the chicks overnight, while the other species did not, and so made appreciably greater provisioning than was recorded during daylight. Certainly, these studies need repeating.

THE IMPORTANCE OF THE GROWTH RATE

The food requirement for body maintenance (producing heat and energy used for body functions, such as digestion and circulation) in kittiwake chicks takes precedence over growth. Figure 6.10 shows an estimate of the amount of food received day by day by a chick that is growing well and also the amount of this that contributes to maintenance. The positive difference between the total and the maintenance is the amount of food available to be converted into tissues, and so resulting in a change of weight. While the precise nature of the graph depends upon the calorific food value of the prey, it is probably a realistic pattern of what occurs with most sources of food used by kittiwakes. The important point is that throughout, most food is used for keeping the chick alive and functioning, while the proportion of food which contributes to the daily change in weight is relatively small. For example, a 10-day-old chick requires about 20g of food per day to produce an increase in weight, and a further 40g for maintenance. If instead of the 60g of food required at this age for an increase of 16g/day, the adults supplied only 58g, that is 2g less, maintenance would still require the same amount (40g), and so only 18g of food remains for growth, which after allowing for assimilation efficiency, allows a reduced daily increase of 14.4g in weight. This shortfall of 2g of an expected 60g of food is a reduction of 3.3% but the growth rate has been

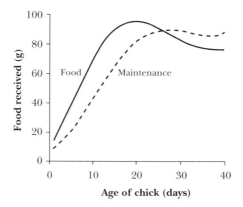

FIG. 6.10   *An estimate of the amount of food received by each chick each day in relation to its age and the amount of this that is used for maintaining the body functions. The difference between the two curves is the amount of food which contributes to growth (or after about 28 days where there is a loss of weight due to the food intake being less than that needed for maintenance). Based on Coulson & Porter (1985), but assuming a higher assimilation rate of about 80%.*

reduced by 10%. The outcome from this and similar calculations is that the daily growth rate of the chick is a very sensitive measure of the food supplied by the parents. It also means that a chick growing at 13g/day is receiving only about 6% less food than another 10-day-old chick growing at 16g/day. Yet evidence suggests that a difference of this magnitude can be critical to the post-fledging survival of the chick.

GROWTH RATES WITHIN A BROOD

Table 6.3 shows that the average daily growth rate in the middle, linear part of the chick's growth varied with brood size and the order of hatching (which invariably followed the order of laying of the eggs). The a-chick (first) in all cases had the highest average growth rate, and in broods of three it was significantly higher than that of the a- or b-chicks in broods of two. The difference probably reflects the greater quality and age of pairs which produced three-egg clutches. The b-chicks grew slightly slower than the a-chicks, while the c-chicks which survived to fledging still had a lower average growth rate than both of the earlier hatching chicks.

Growth rate would appear to have an important effect on the subsequent post-fledging survival during the following two years (Table 6.4). This data set used the growth rate of chicks of those individuals which survived at least two years and were subsequently seen as potential breeders in the colony at North Shields. The results show that when the growth rate of chicks was higher, so was the proportion that survived to return there at least two years later. The graph of these results (Figure 6.11) shows that the points closely fit the straight line of best fit. Extrapolation of the line to the point when no chicks returned two years later suggests that chicks which did not exceed about 7g per day in their growth failed to survive to adulthood. A word of caution is needed here. Many of the surviving chicks from North Shields did not return to their natal colony but bred elsewhere. However, it has been assumed that the sample which returned to North Shields was typical of all of those which survived, but this assumption cannot be confirmed with certainty. Some of these were found in other colonies, but none (n = 25) had growth rates below 12g/day; the average was 15.8g/day.

TABLE 6.4 *The proportion of kittiwakes which were later seen at North Shields when two or more years old according to their maximum growth rate as chicks.*

Growth rate (g/day)	<12.0	12.0–15.9	16.0–16.9	17.0–18.9	Over 18.9	Total
Number that fledged	62	385	257	167	33	904
Number that returned as adults	4	33	28	24	5	94
Percentage that returned	6.5	8.6	10.9	14.4	15.2	10.4

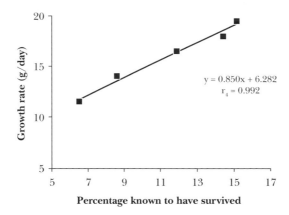

FIG. 6.11 *The relationship between the daily growth rate of individual kittiwake chicks between 75g and 300g and the percentage which subsequently returned to the colony more than two years later. It has not been possible to include individuals that moved to another colony, and did not visit North Shields.*

DOES THE SUCCESS OF DIFFERENT CLUTCH SIZES REFLECT THE QUALITY OF INDIVIDUALS?

It was argued by David Lack (1954) that the most frequent clutch size tends to be the most productive, and he supplied considerable evidence in support of this, although several exceptions have been reported (e.g. Roff 1992) suggesting that birds often lay smaller clutches than might seem optimal. The kittiwake is one such exception. Although two eggs was by far the commonest clutch size at North Shields (and in most other colonies), on average, each clutch of three eggs produced 40% more fledged young than each clutch of two eggs, while the production of fledged young per egg was 53% higher than from two-egg clutches. The productivity of single-egg clutches unexpectedly had a much lower success rate per egg than in the larger clutches, and 67% of the eggs failed to produce a fledged chick (Table 6.5).

It is possible that the dismal performance of pairs laying a single egg was a result of many of these birds breeding for the first time, as they typically had an appreciably reduced breeding performance. However, when all first-time

TABLE 6.5 *The clutch size and production of young to fledging by kittiwakes at North Shields, 1954–90. Based on 2,661 clutches.*

Clutch size	Number of clutches	Percentage of all clutches	Young fledged per clutch	Percentage success per egg to fledging	Percentage of clutches that were total failures
One egg	275	10.3%	0.33	33%	67%
Two eggs	2070	77.8%	1.14	57%	22%
Three eggs	316	11.9%	1.74	58%	10%

TABLE 6.6 *The clutch size and production of young to fledging by kittiwakes at North Shields, 1954–1990, according to clutch size. Based on 1,466 clutches where the male and female of each pair were both breeding for at least the second time.*

Clutch size	Number of clutches	Percentage of all clutches	Young fledged per clutch	Percentage success per egg to fledging	Percentage of clutches that were total failures
One egg	81	5.5%	0.46	46%	53%
Two eggs	1130	77.1%	1.31	66%	19%
Three eggs	255	17.4%	1.84	61%	11%

breeders of either sex are excluded from consideration (Table 6.6), the overall success rate per egg increased for all clutch sizes, but the lower success rate arising from single-egg clutches was still 30% lower than that from two-egg clutches and it is obvious that factors other than first-time breeding are involved and result in a total failure rate of 53%. When breeding success is broken down into the egg stage (hatching success) and chick stage (fledging success) (Table 6.7), the poor performance of single-egg clutches is totally attributable to a failure during incubation, while if the egg hatched, the fledging success was then comparable to that in larger clutches. Predation did not play a part in this failure during incubation, nor were there many addled eggs among these. In most cases, incubation ceased before hatching could occur.

At North Shields, kittiwakes that produced three-egg clutches were appreciably more productive than those with smaller clutch sizes, despite the parents usually having more young to feed. The percentage hatching success per egg from two- and three-egg clutches did not differ meaningfully, while the survival to fledging of the young that hatched, measured by fledging success, was similar and not meaningfully different between any of the three clutch sizes (Table 6.7). While the overall percentage success of three-egg clutches was slightly lower than that in clutches of two eggs, the difference is not meaningful and despite the large samples it is not possible to conclude that there was a difference in breeding success per chick hatched between those produced from two-egg and three-egg clutch sizes.

These results are surprising in several ways. They demonstrate that under the conditions at North Shields, kittiwakes laying three eggs were much more productive than those laying two eggs. It has already been shown that the first two chicks from broods of three had, on average, slightly higher growth rates than the two chicks from two-egg clutches, so there is no reason to suspect that their subsequent survival was inferior. The fact that the parents had an extra chick to feed in broods of three chicks did not detract from their ability to rear their young. This indicates that the parents did not frequently encounter difficulty in obtaining the extra food required by a larger brood. Accordingly, this situation did not comply with David Lack's hypothesis that

TABLE 6.7 *The hatching and fledging success of male and female kittiwakes that had bred before in relation to the clutch size.*

Clutch size	Number of clutches	Percentage hatching success	Percentage of young hatched that fledged	Percentage breeding success
One egg	81	53%	88%	46%
Two eggs	1130	75%	88%	66%
Three eggs	255	73%	84%	61%

the most frequent clutch size was also the most productive, nor did it fit with more recent modifications of this relationship (Charnov & Krebs 1974).

There is no evidence in this study of the kittiwake that laying an extra egg above the normal clutch of two impinged on the parents' ability to raise their brood. This result is in conflict with the findings of Monaghan *et al.* (1998) who found that Lesser Black-backed Gulls which laid an extra egg were less successful in rearing a normal brood and so, apparently, the pair incurred a cost simply by the production of an additional egg.

In passing, it is worth noting that both the males and females of pairs which laid three-egg clutches in a particular year actually had higher (but not meaningfully so) survival rates in the following 12 months than those laying two eggs. These results for the kittiwake strongly suggest that laying an extra egg and rearing an extra chick did not burden the parents with detectable costs affecting their survival or their ability to rear the brood. Perhaps the main difference between the experiment by Monaghan *et al.* (1998) and our study was that they induced female gulls to lay an extra egg (by removing one egg already laid) in addition to those which the birds would naturally have laid. In the North Shields kittiwakes, the females had 'decided' to lay the larger clutch of three eggs. A parsimonious explanation would be that many birds lay a clutch which reflects their quality and that those kittiwakes which laid three eggs were better-quality individuals than those laying two-egg clutches, 'knew' they were better, and were able to accomplish the extra work involved by laying an extra egg without a detectable cost.

If there was an inherited tendency to lay a particular clutch size, then under the conditions at North Shields, there should have been an increase in the proportion of females laying three-egg clutches, since such genes would be expected to become an increasing component of the population. This trend was not detected, but it should be remembered that most of the female kittiwakes breeding at North Shields had been reared in other colonies (Coulson & Coulson 2008), and may have come from colonies where the success rates from different clutch sizes were different from those at North Shields.

I have noted that in the majority of cases where a pair of kittiwakes fledged three young, one of the well-grown chicks could move off the nest and onto part of the ledge on which the nest was built. In this situation, there was still room from time to time for two adults and three chicks to be present at the nest site while still allowing the chicks to exercise their wings. But this was not possible at all nest sites. So the success of large clutch sizes may be influenced by the type of site the adults could use. If the nest is not on a longer ledge, or if there are other nests immediately alongside, it is probably much more diffi-cult for kittiwakes to successfully rear all three chicks. But this problem has already been included within the data previously given. There is also the possibility that an abundant food supply has only become available to breeding kittiwakes recently and they have not had sufficient time for evolu-tion and selection to take advantage of this. It is difficult to comment further on this, other than to point out that food seems to have been abundant in Britain since the start of the 20th century or earlier, as indicated by the popu-lation growth, and this may represent the passing of more than 15 generations. Is this time enough for selection to have occurred?

There is another possibility, and one which I have already touched upon, namely that clutch size is not rigidly determined genetically, but that the number of eggs is modified by the condition of the female (or even the male). In effect, kittiwakes are producing the clutch size which is appropriate for their individual condition or 'quality'. It is evident that many of the kittiwakes laying a single-egg clutch are poorer-quality individuals and they are not only young, inexperienced birds. If this was not so, the incubation of the single egg by experienced adults would have been more effective than is indicated by the value shown in Table 6.7. If appreciable differences in quality do exist between individuals, the experiment which has been used on several occa-sions of increasing or reducing the clutch size of individual pairs and recording the outcome would not be valid, because care was not taken to compare like with like in the control and experiment. It would be necessary to measure the quality or condition of the birds being used in the experiment in advance. Can this be done?

It is possible to construct a hypothesis that pairs of good-quality individuals lay early, and it is known that early-laying birds are much more likely to lay three-egg clutches. Individuals whose quality has deteriorated, either because they are ill or for other reasons, will lay later. As shown in Chapter 5, this leads to a tendency, even in the same individual, to lay fewer eggs. In effect, the best individuals in the colony are those nesting early. This even applies to first-time breeders, where success declined if they laid late in the breeding season. The problem with this hypothesis is how to measure 'quality' and then test the hypothesis without circularity invalidating the interpretation of the outcome. There may be much evidence that the quality of individuals varies, but how should this quality be measured and quantified?

Currently, we know little about the variation in quality of individuals or how to quantify this. To what extent is quality measured by the length of life of an individual, since this is the key factor in determining the life-time breeding success? In all probability, the length of survival is only partially dependent on quality, because some causes of mortality are the result of chance effects, such as being killed by a Peregrine Falcon while flying over the sea, or stones falling from higher up the cliff and killing an individual on a nest.

Are longer-lived females of better quality? An insight into this question can be obtained by comparing the number of young produced each year on their first or second breeding occasions by females which survived to breed only once with the production of young in the same first year of breeding by individuals which survived longer.

The results of this analysis are shown in Figure 6.12 and the data have been considered separately for the centre and edge of the colony to eliminate possible effects stemming from position in the colony. Considering only the first two breeding occasions, centre females that did not survive for more than two years after starting to breed fledged 33% fewer young per year during this period than those females that survived and went on to breed on more occasions. At the edge of the colony, the reduction was about 20%.

These results suggest that on average, poorer quality and shorter life span are linked, but there was no indication that after breeding for three years, the presumed greater quality of the longer-lived individuals was indicated by the level of their earlier breeding performances.

Is the position of a nesting bird in the colony (the centre and edge effect) a measure of quality? Does the poorer breeding performance in the year before death, which occurs in kittiwakes, indicate a deterioration in quality? How long is good quality retained? I suspect that size does not play a part in determining quality in kittiwakes, since some of the most productive breeders in the colony were relatively small individuals. The concept of quality variation in individual kittiwakes was originally advanced by Coulson & Porter (1985), although the general idea is, of course, embedded in natural selection and evolutionary theory. The concept of differences in quality is attractive, but it is difficult to quantify for specific individuals

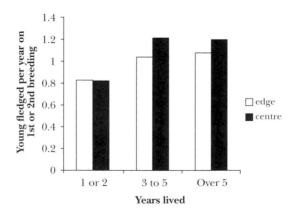

FIG. 6.12 *The average number of kittiwake chicks fledged per pair each year during their first and second breeding attempts at the centre and edge of the colony and according to whether the adult female survived for only one or two breeding years (n = 99 edge and 68 centre), for three to five years (n = 72 and 78) or for more than five years (n = 127 and 128). Data from females starting to breed from 1960 to 1980 at North Shields.*

while they are still alive. Yet a means of measuring quality is necessary before it is possible to conduct meaningful field experiments, such as those varying clutch and brood sizes.

PRODUCTIVITY IN BRITAIN, IRELAND AND ELSEWHERE

Since 1986, the Joint Nature Conservation Committee (JNCC) has acted as a depository of seabird census data collected around the British Isles, mainly by interested amateurs and wardens of nature reserves. These data have been collected annually for a number of kittiwake colonies and groups of colonies around the British Isles and I am most grateful to the staff of JNCC and the many individuals who collected data to have been able to examine and analyse this extensive data set. Not only are regular counts made of the size of colonies, but information is collected on the annual average productivity of individual pairs, and it is these data that are considered below. Data used in this chapter are listed in Appendix 3.

From about 1987, reports of very poor breeding success started to be recorded in Britain, and questions began to be asked as to whether kittiwake productivity was high enough to sustain the current national population. To make a judgement on this, it is necessary to estimate the production of young per pair needed to sustain numbers, and an approximate and provisional estimate is made below.

If adult kittiwakes suffer an annual mortality rate of 15% (an average for European data; see Chapter 10), then to ensure no decline in the colony size of 100 pairs, 30 new breeding individuals are needed. The survival rate from fledging to breeding was estimated at 36% (see Chapter 10), so the 30 recruits would be the survivors of 83 young which fledged, that is, 0.83 young per nest.

This is a rough calculation, but it suggests that about 0.8 young need to be fledged per nest each year if the overall production is going to replace the deaths of adults and so maintain numbers. Thus it is not unreasonable to take an average of 0.8 young per pair (the critical level) as a guide of reproductive success needed by kittiwakes in Western Europe without a population change. Based on the estimates of adult survival at Middleton Island in the Pacific, a level about half this is necessary, because of the high survival rates of adults in this area.

In the years leading up to 1986, few, if any, cases of poor breeding by kittiwakes were reported from the British Isles, although years of low breeding success had been reported from other North Atlantic colonies from time to time, such as at Røst in Norway. However, it is not certain that breeding failures did not occur in British colonies during that time, because observations on colonies were much less extensive and records of breeding success year by year were rarely maintained.

The only long-term data run which exists is from 1953 to 2010 and was collected in the Tyne area of north-east England, but more recent data exist for other areas.

Figure 6.13 shows the production of young per nest from three areas. At one extreme, Kittiwakes in the Tyne area of north-east England had 45 successive years of high breeding success (Figure 6.13A). This declined below the 0.8 young per nest only in the period of exceptional adult mortality in 1998 and 1999, which killed many kittiwakes while they were rearing their young. Thereafter productivity recovered and has mainly been above the critical level. There can be little doubt that this area has been producing more than the necessary number of young each year for 56 years.

Figure 6.13B shows a series of data for Foula (collected by R.W. Furness) and is separated from information from the rest of Shetland because of the island's isolation. From 1971 to 1985, Foula kittiwakes had a consistently high breeding success, but thereafter productivity there and elsewhere in Shetland declined and showed considerable fluctuations from year to year. Foula returned to a high productivity, so that between 1992 and 2000 it dropped below the critical 0.8 level in only two years, but in the rest of Shetland there was only one year between 1986 and 2000 in which productivity exceeded the critical level. Since 2000 both areas have failed to reach the critical level in any year.

Figure 6.13C shows the data for the Isle of May. Between 1985 and 2005, productivity reached the critical level in only six widely spaced years. The contrast between these areas is considerable. The correlation between the productivity on Foula and the rest of Shetland (Figure 6.14A) is high, but Foula shows more year-to-year variation. In contrast, there is no correlation whatsoever between the productivity in Shetland and on the Isle of May (Figure 6.14B); their good and bad years were different.

There was a modest correlation between the Isle of May and the Tyne area (Figure 6.14C) and while they often shared good years, the coincidence of poor years was less frequent.

It is evident that bad breeding years were often local and, in many cases, were not caused by widespread food shortage.

Table 6.8 considers 19 colonies for which there is productivity data for over 20 years between 1986 and 2009. A slight majority of these had over 50% of years with 0.8 or more young fledged per pair. Productivity was high along the east coast of England and as far north as St Abbs in Scotland. In contrast, productivity was low in colonies adjacent to the Irish Sea, but high on the rest of the west coast of Scotland. The difference between Orkney and Shetland is marked, while St Kilda, to the west of the Hebrides, had the poorest performance.

This aspect can be investigated further by examining patterns of productivity in a sample of 16 colonies or areas where data were available around the British Isles, but I excluded data from the Tyne area because it showed minimal variation and was also close to the Farne Islands, which was used. Figure 6.15 shows the average number of young fledged per pair in the 16

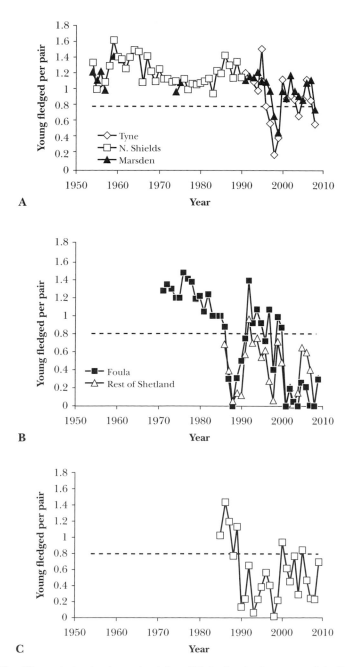

FIG. 6.13   *The variation in the productivity of kittiwakes in three areas of the British Isles.* **A.** *The Tyne area in north-east England (North Shields, Marsden; author's data) and other colonies on the River Tyne (D. Turner in litt.).* **B.** *Foula and the remaining areas of Shetland (JNCC data)* **C.** *The Isle of May in south-east Scotland (JNCC data). The horizontal dashed line indicates 0.8 young fledged per pair, which is an estimate of that needed to maintain the population.*

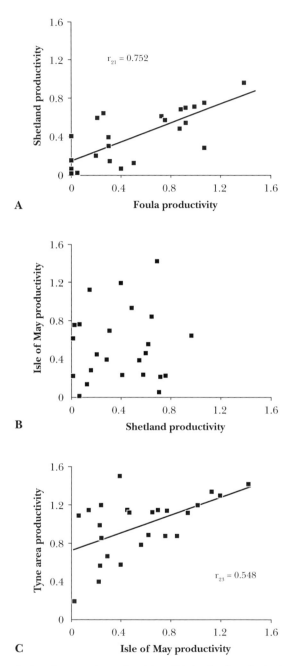

FIG. 6.14 *Correlations between the productivity of kittiwakes in colonies in the same years from 1987 to 2009.* **A**. *Foula and the rest of Shetland,* **B**. *Isle of May and Shetland,* **C**. *Isle of May and the Tyne area. There was no indication of a significant correlation between Shetland and the Isle of May but there was a high correlation between Foula and the other Shetland colonies and a modest but significant correlation between the Isle of May and the Tyne area.*

TABLE 6.8 *The percentage of years between 1986 and 2009 in which the productivity in the colony was 0.8 young per nest or higher, which is considered satisfactory. The colonies are listed from north to south. Colonies with low productivity are shown in bold type. It takes 50–60% of occasions with 0.8 or more young per pair to average 0.8 young per pair each year (the critical productivity) over the whole period.*

East coast colonies	Percentage	Northern colonies	Percentage	West coast colonies	Percentage
N. Sutor, Nigg	30%	**St Kilda**	4%	Handa	83%
Isle of May	17%	**Shetland**	12%	Canna	62%
St Abbs	64%	**Foula**	37%	**Ailsa Craig**	17%
Tyne area	87%	Fair Isle	58%	**Skomer**	29%
Farnes	52%	Orkney	70%	**Dunmore East**	37%
Saltburn	57%				
Bempton	70%				
Lowestoft	83%				

colonies from around the British Isles since 1986, when extensive recording was started. Surprisingly, from 1986 to about 2000 the average productivity remained virtually constant and slightly above the estimated 0.8 young threshold level required for stability, despite several records of breeding failure during this time in individual colonies. The reason for this constancy was that the years of failures did not coincide, again suggesting that they were local effects.

So during the 20th century, there was little to be concerned about in regard to the national productivity of kittiwakes. All of the evidence suggests that it remained high throughout, although by the end of this period the productivity was probably no longer high enough to sustain the population growth which was evident over much of the 20th century.

However, this conclusion is at odds with the national census of breeding pairs of kittiwakes made during Seabird 2000, which reported a decline of 23% in Britain and Ireland from 1986 to 2000, which is a decrease of about 2.0% per year, and not 2.3% as stated in the kittiwake chapter in Mitchell *et al.* (2004). The census of the 23 major colonies in Britain and Ireland in 2000, presented in their Table 2, revealed a decline in 15 of these when compared with the census in 1985–88. Examining changes identified by the entire census, Shetland showed the greatest regional change with a decline of 69% over this period or a decrease of 7.5% per year. The next largest decline was reported from north-east England of 40% over the same time period (–3.3% per year), but as argued in Chapter 11, the census on the large Bempton-Flamborough complex which contributed to the baseline for measuring this decline was probably appreciably exaggerated and, if this is accepted, then the actual decline in the area was marginal at 5.5% over about 15 years or about 0.4% per year. In 2000, the numbers in north-east England were still appreciably higher than those for 1970.

The next largest decrease reported was in south-east England, where there

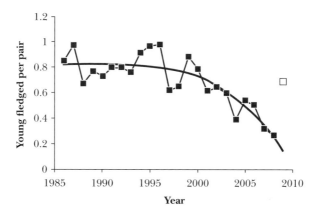

FIG. 6.15 *The average production of fledged young from a sample of 16 kittiwake colonies around the British Isles from 1986 to 2009. The trend line is a third-order polynomial fitted to the data from 1986 to 2008. Data for 2009 is shown as an unfilled square and is very much higher than the trend from previous years. Data from the Tyne area, where productivity was high throughout the period under consideration, is not included. The colonies or areas included are as follows.* Scotland: *Canna, Ailsa Craig, Colonsay, Handa, Orkney, Shetland, Fair Isle, St Kilda, North Sutor, St Abbs.* England: *Farnes, Saltburn, Bempton, Lowestoft.* Wales: *Skomer.* Ireland: *Dunmore East.*

are only a few thousand pairs and a decrease of 34.8% was recorded. I cannot reproduce this figure and, assuming that the Isle of Wight was included in the south-east region, then the decrease was 26% or a loss of 2.2% per year.

The next highest decrease between the 1985–88 and the 2000 census was in north-west Scotland (Table 3 in Mitchell *et al.* 2004) with a decline of 24.8%. Unfortunately, the geographical limits of this region were not defined, but for every administrative area within the region, the decline of the kittiwake shown in their Table 1 was less than this! So presumably, there can be little confidence in the stated 23% decline in the kittiwake in Britain and Ireland between 1985–88 and 2000. Without doubt, there had been a marked decline in Shetland and a smaller decline in Orkney. Elsewhere, the decline up to 2000 has often been exaggerated.

Productivity of the kittiwake in Britain did change dramatically from 2001 to 2008, with the average annual production of young progressively declining so that, from 2005 to 2008, it was reduced to half the earlier values, with an average of only 0.4 young produced per pair in the sample. This was a more significant indication of a potential major threat to kittiwake numbers, and was just reason for despondency as the trend line progressively became lower. However, in 2009, breeding success at most colonies abruptly increased markedly, and the average nearly reached the critical level of 0.8 young per nest (and preliminary figures suggest that this improvement was maintained in 2010).

It should be noted that because of the length of the immature period in the kittiwake, the overall lower productivity that started in 2001 could not impact on the numbers of breeding birds until 2005 at the earliest. The conclusion is that up to 2000, the overall productivity of the kittiwake in Britain gave no major cause for concern, and if a major decline in the numbers of breeding kittiwakes in Britain had taken place by then, it had to be attributed to an increase in the mortality rate of fully grown individuals.

The reasons for the lower productivity between 2000 and 2009 are not known, but a reduction in available food over larger areas appeared to be implicated. A few colonies suffered heavy predation at nests by Great Skuas and other avian predators, and this also may have been initiated by food short-ages for the predators. The key food organism is the sandeel and knowledge of sandeel numbers and distribution are still not well known or understood, but the foolish exploitation of this fish for non-human use could have major impacts on many marine animals which use it as food, such as seals, many seabird species and larger fish like Cod. The harvesting of sandeels should be severely restricted, irrespective of whether this species has a major impact on the kittiwake. Whether the improvement of kittiwake productivity in 2009 and 2010 is the beginning of a major recovery in the breeding success or has been just a 'flash in the pan' remains to be determined in the future.

While there is an indication of a decline in the size of colonies in areas where breeding success has been low, e.g. Shetland, in general, the correla-tion between the two is relatively poor. The main reason for this is the low level of philopatry in the kittiwake, and the influx of new breeders reared in colonies elsewhere is important. For example, the assumption that the low productivity of kittiwakes on Fair Isle will cause the colony to disappear in the next few years (Rothery *et al.* 2002) is too simplistic a view of the population dynamics of the species and ignores the importance of immigration. That is not to say that philopatric individuals are not important, but the well-being of a colony is also affected by the production of young elsewhere, possibly within a radius of several hundred kilometres. Any doubt about the importance of this is surely dispelled by the way young colonies grow appreciably in the first few years of their existence and this is entirely caused by immigration (see Chapter 11).

Productivity elsewhere

There are few areas within the distribution of the kittiwake where the produc-tion of young has been monitored regularly for a number of successive years. Such information has been collected by Barrett & Tertitski (2000) for northern Norway and Russia, and numbers of years with very low productivity have been recorded and associated with crashes in Capelin stocks. Elsewhere in Norway, years of breeding failure have also been reported, some going back 50 years, but regular monitoring has started there only in the last few years. Similarly, poor breeding years have been reported sporadically in

Newfoundland and Labrador (Chapdelaine & Brousseau 1989, Birkhead & Nettleship 1988), again associated with fluctuations in Capelin.

In the Pacific Ocean, Hatch *et al.* (1993b) have documented the poor breeding success of kittiwakes breeding on Middleton Island, Alaska, and the experimental supplying of supplementary food to chicks there (Gill & Hatch 2002) clearly demonstrated that the normally low success could be attributed to limitations in the amount of food that could be collected by adults. Elsewhere in the Pacific, poor breeding years have been interspersed with occasional years of greater success (Gulf of Alaska and Bering Sea, Baird & Gould 1983; Pribilof Islands, Dragoo & Sundseth 1993; Bering Sea, Biderman *et al.* 1978, Murphy *et al.* 1991) and marked differences have been found in the same year between colonies (Suryan & Irons 2001), but with most having productivity far too low unless adult survival rates were exceptionally high.

The breeding success of the Red-legged Kittiwake has been summarised by Byrd & Williams (1993) for the period 1973 to 1992. Overall, 0.43 young were fledged per pair based on 28 estimates (some years with more than one estimate, but from different colonies) and this figure did not take into account variable proportions of adults which did not lay. Total failures were recorded in some years.

YEARS OF FAILED BREEDING

The occurrence of years when the breeding success of kittiwakes is very low or failed has produced expressions of concern for the existence of the colonies, and even for the species. Such concerns go back many years to the failed breeding on Røst, in the Lofoten group of islands off the coast of Norway in the 1950s. Often, these are based on the erroneous belief that the colony has to produce enough young to ensure its own continuation.

However, little consideration has been given to the possibility that a year (or multiple years) of failed breeding or low productivity in individual colonies might be the norm for the kittiwake, and that concerns might be unjustified. In fact, the high breeding success of kittiwakes in Britain in the 20th century may well be the exception, produced by the recovery from persecution and a favourable environment, allowing years of successful breeding. This exceptional situation of continuous, successful years has existed from 1953 to date at North Shields and Marsden in north-east England, where breeding success remained high except for one brief period when a toxin affected the adults during the breeding season.

The method of feeding used by kittiwakes (and sea terns) – capturing food in the surface waters of the sea – is a high risk strategy and seems to be precisely the system that is likely to generate year-to-year variation in breeding productivity, and so we should not be surprised when poor breeding years occur. The local nature of poor years suggests that the cause is related to the local availability of sandeels or Capelin, and the lack of alternative fish species that can be exploited. In some years and places, there may have been wide-spread crashes in stocks of food fish for kittiwakes, as has happened in Norway and Newfoundland, and these took several years to recover. A consequence of this is often a series of years with poor breeding success by kittiwakes, which continues until the fish stocks recover. Barrett & Krasnov (1996) and Barrett (2007b) gave excellent examples of this effect in northern Norway and north-west Russia.

In Britain, years of low productivity have not yet caused the loss of individual kittiwake colonies, although some have decreased in size. This can be readily explained firstly by the relatively low level of failures up to 2000, which did not appreciably decrease nationwide productivity to critical levels, and secondly by the recent discovery that much of the recruitment of new breeding kittiwakes comes from other colonies, while relatively few return to breed in their natal colony. These factors result in the effect of a year with local poor breeding being diluted and spread over a wide area.

In Britain, an exaggeration of the decline in kittiwake numbers in recent years, and the possibility that the occurrence of some poor breeding years might be normal, suggests that there may have been an over-reaction to the situation up to 2000. Since then, declines in numbers have continued in the northern islands of Britain, but elsewhere the situation seems not to have changed dramatically, although average breeding productivity declined between 2001 and 2008. However, the first effects of this decline on colony size could not be expected before 2005, when the surviving young that fledged in 2001 would start breeding for the first time, and even by 2011, the effect of this perturbation would not have reached its peak. Firm conclusions on the impact of reduced breeding performance at the start of the 21st century must await the next national census of colonies. In the meantime, the placement of the kittiwake, the most numerous gull in the world, as an 'amber' endangered species in Britain may have been premature, perhaps based on errors in census work, a failure to understand alternative implications which can be drawn about breeding failures which have occurred occasionally or even in several successive years, and a failure to appreciate the implications arising from the huge increase in kittiwake numbers during the 20th century.

The breeding success of kittiwakes is clearly related to the stocks of a few fish species, particularly sandeels and Capelin, leading to the question as to why the numbers of these fish species change over time. One possibility is that these are natural fluctuations in response to varying natural environmental conditions (Murphy *et al.* 1991), such as those affecting the success of their spawning. Another possibility is the effect of human activity through the over-exploitation of fish. The evidence of human influence is not strong, even on a local scale. It remains to be proven definitively that human exploitation of sandeels has widely influenced kittiwake breeding success and caused numbers to decline, although exploiting Capelin and sandeels for non-human consumption is clearly a folly.

No doubt there will be attempts to associate the variations in breeding success of the kittiwake with climate change. It may be that this will become important in the future, but currently, the changes in average sea temperatures are still small (about 1°C in the past 30 years) and, since year-to-year fluctuations are much greater, it is difficult (for me, at least) to believe that these have already appreciably influenced the kittiwake throughout its whole range. That is not to say that changes have not occurred in the marine ecosystem, but the demonstration of effects on kittiwakes is speculative and not yet well founded.

AFTER THE YOUNG FLEDGE: THE ANNUAL DESERTION OF THE COLONY

Immediately after the chicks leave the nest for the last time and head out alone to the open sea, the parent–chick bond is broken, the presence of the parents at the nest becomes irregular and often the nest site is left unoccupied overnight. One or both parents return to the colony each day, usually in the morning, and then often depart and return several times during the day, but no clear pattern of visits is evident and as a consequence the site is often left unoccupied. However, there is no intense competition for nesting sites at this time as the last prospecting sub-adults ceased to visit the colony a week or more earlier. The continued pattern of attendance of adults at the colony seems to depend upon the presence of unfledged chicks in some of the nests and the time when the last chicks fledge is often, but not always, within a few days of the date on which the whole colony is deserted for the rest of the year.

In any particular geographical area, the last chicks to fledge do so at a similar date in all colonies. This is usually in the first week of August in Britain, but a month later in Alaska and the higher Arctic areas of the North Atlantic (Nettleship 1974). As a result, the normal annual vacation of the colony tends to be later further north. Along the coasts of Devon Island, in the North West Territories of arctic Canada, the autumn departure of kittiwakes from the whole area, including colonies, is abrupt and McLaren & Renaud (1982), using aerial surveys, counted 24,000 kittiwakes on 25 September 1976, but these were reduced to 5,500 on 29 September and only 250 kittiwakes remained on 5 October 1976.

However, in some regions the behaviour of the adults when the last chicks leave the colony can be very variable, and this was evident between years even in the same colony. For example, in several years in north-east England, the last chicks fledged during the first week in August but many of the adults returned daily to the nest sites in the colony until the end of the first week in November, some 13 weeks after the last chick fledged. Yet in most other years, the adults vacated this colony at the same time as the last chick in the colony finally fledged. In one extreme year, they left six chicks on nests after all of the adults had left the colony for the last time in that year. These chicks were near fledging and then left within two days.

The variation in the behaviour of the adults at the end of the breeding season appears to be associated with the presence or absence of panic flights. In most years, the onset of the vacation of the colony is accompanied by panic flights, a behaviour similar to that seen in the early stages of the annual reoccupation in the spring and when there seems to be a conflict developing between attending the colony or returning to a totally pelagic life. At the end of the breeding season the massed panic flights and repeated return to the cliffs affected many and sometimes all of the adults present, and occurred with increasing frequency. As in the early spring, these panic flights are 'vacuum' activities and are not triggered by an external stimulus, such as the presence of a predator. Characteristically, calling within the colony suddenly stops and there are two or three seconds of total silence. The adults become alert and wary, standing up and looking around. Suddenly, one bird leaves the nest in a steep dive and is immediately followed by most or all of the individuals from a considerable part of the colony pouring off the cliffs. Sometimes, all of the birds present on over 600m of cliff left almost synchronously, while at other times, only a smaller part of the colony is affected. As in the spring, a proportion of the birds return to the cliffs after a few minutes, only for further panic flights to occur, eventually resulting in a total evacuation of the cliffs.

Following each panic flight, some birds leave and fly directly out to sea with determined flight and in an unvarying direction towards and beyond the horizon, typically emptying the colony during the early afternoon. On the next morning, a decreased number of birds return to the colony. More panic flights occur on each of several successive days and these result in the colony being emptied of adults by midday. After panic flights have occurred on several days, no birds return on the following morning and the colony is deserted until the spring. The desertion of the colony each year closely follows the pattern of the reoccupation in the spring, but in the reverse order.

In years when adults remained at the colony for several weeks or even months after the last chick fledged, intense panic flights did not occur. Non-breeding prospectors no longer visited the colony, nor did adults which had bred for the first time that year, and the continued occupation involved only the older adults. The colony was deserted each night and the period of daylight when birds were present, as well as the maximum numbers, progressively became less and less, centred on 10.30 hrs. However, the daily departure of individuals used the normal shallow dive from the nest and not the steep dives associated with panic flights. The last birds to attend the colony were of both sexes and were almost entirely the older individuals which had bred on many occasions, and many remained in pairs.

What causes the annual differences in departure dates? This may be linked with calm weather and good feeding conditions within easy range of the colony, with no advantage gained by leaving the colony to reach better feeding grounds further away, but evidence supporting this hypothesis does not exist.

Years when some adults remained at the colony up to three months after the last chick fledged coincided with an early reoccupation of the colony by the older birds in the following spring, and in these years some of the adults were absent from the colony for only two months. In general the first to leave in autumn were the last to return in the next spring. Does the date of departure in the autumn influence the time of return in the spring, with the later they leave the earlier the spring return? I suspect this is so and it also applies to failed breeders that typically left the colony early.

CHAPTER 7

Weights of kittiwakes during breeding

KITTIWAKES differ in weight and general body size over their geographical range, and males are larger than females. In addition, there are also consistent changes in the weight of individuals during the breeding season. During the study at North Shields, north-east England, each kittiwake captured for marking was weighed at the same time, and body measurements were taken. Over the years, more than 1,500 adult and prospecting kittiwakes were weighed, mainly in the period from May to July, with a few weighed earlier or later in the breeding season. Almost all of these individuals were colour-ringed and sexed, and all were known as either non-breeding prospectors or breeding individuals in the year of weighing.

Analysis of these weights gave an insight into the changes that occurred during the breeding season for both sexes and for breeding birds and potential recruits. As shown in Chapter 9, the annual average weights did not change appreciably or systematically during the study, and so the data for all individuals have been pooled and their weight examined through the breeding season.

In several avian studies, loss in weight while rearing young has been regarded as an indication of the cost to the individual of breeding (e.g. Golet & Irons 1999, for kittiwakes). In other studies, weight, often corrected for structural size of the individual, has been used as a measure of individual quality.

A little reflection raises concerns about making these assumptions. Weight may well be an indicator of the quantity of food reserves, such as fat deposits, but it is also a measure of the amount of effort that has to be employed during flapping flight, where the heavier individual must expend a greater amount of energy to remain airborne, thus counteracting the effect of gravity. So while it may be important for a bird to have food reserves, they can also present a disadvantage by requiring more effort during flight. There is probably a compromise between the two needs, but this could change with season or when the amount of daily flying changes. Food is more likely to be sparse in winter, and more abundant and predictable in the breeding season. Thus weight of an individual is likely to be greater in winter as this means more food reserves which contribute to survival in severe weather. In contrast, many species rearing young spend more time flying to obtain the additional food required for chicks. Moe *et al.* (2002) found that breeding kittiwakes at Svalbard (79°N) in one breeding season decreased their weight in the first part of the chick-rearing period, with females losing 14.8% and males 8.4% of their body weights, and then remained at those levels until the young fledged.

I was able to examine weight changes in kittiwakes during the breeding seasons at North Shields, using the extensive series of weights gathered over many years. The date of egg laying for each breeding individual was known in the same year that the weight was recorded, and the two values could be connected because each bird was colour-ringed and identifiable.

When the weights of female kittiwakes at North Shields were examined in relation to the date of egg laying and hatching, there was an increase in weight just before eggs were laid, presumably caused by the growth of developing eggs within the female. Following the completion of the clutch, there was a prompt return to the pre-laying weight. The weight of both males and females slightly increased during incubation. However, when the eggs hatched, the weights of both sexes dropped by 6% within only three days, and remained at this level until after the young had fledged some six weeks later (Figures 7.1 and 7.2). This pattern was similar to that reported by Moe *et al.* (2002), but the weight loss at North Shields among females was appreciably less.

There are two important points regarding this effect. First, my data were spread over more than 30 years and, when considered over fewer years, the pattern persisted in all of these, indicating that the drop in weight was a

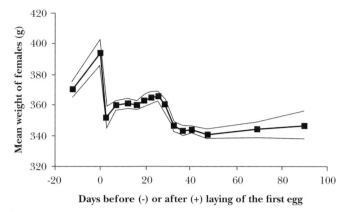

FIG. 7.1 *The mean weights of breeding female kittiwakes during the breeding season. The eggs hatch on or about day 27, when there is a sudden drop in weight. The thin lines indicate one standard error from the trend line. After Coulson (2010).*

regular feature and not an effect that occurred in occasional years. Second, this decline occurred long before the food demands of the young became appreciable and clearly was not a response to an immediate need to collect more food with increased effort involved. Females remained at this lower weight level, and did not decline further at the time (after 10 days) when the food needs of the chicks became greatest.

I remain puzzled by the claim of Moe *et al.* (2002) that the early chick stage is the time at which energy requirements of adult kittiwakes are greatest, because at this time the food needs of the chicks are small. I do not understand how greater energy expenditure is required at this stage, because brooding just replaces the attendance of a parent during incubation, and the

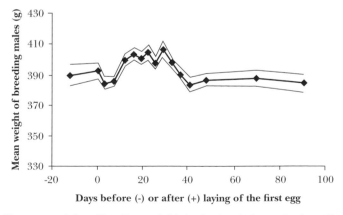

FIG. 7.2 *The mean weights of breeding male kittiwakes in relation to the time of breeding of the pair. The eggs hatch on or about day 27 when there was a progressive decline in weight of the males. The thin lines indicate one standard error from the trend line. After Coulson (2010).*

adults' off-duty time remains the same. Very little extra flying is needed by the off-duty parent to obtain the food required by small chicks. Further, Moe *et al.* claimed that breeding effort was sex-dependent, because their measured weights indicated that females lost more weight than males after the eggs hatched. However, their data do not seem to justify this contention and I can find no meaningful significance in the differences of weight losses between the sexes. It may be that the authors have put too much confidence in the accuracy of their own and others' estimates of energy requirements, particularly as no confidence intervals were presented in their earlier paper (Fyhn *et al.* 2001) to indicate their accuracy.

Another explanation for the sudden drop in weight is that it increases the efficiency of flying associated with food collection for the young. To feed two well-grown chicks, each parent probably has to increase its total flying time by 20–30%, and a drop in body weight is an effective strategy to conserve energy. What is surprising is that this loss of weight occurred so rapidly. This idea is not new and in seabirds, Gaston & Perin (1993) found a similar loss of weight at hatching and suggested the same explanation for Brünnich's Guillemots. More recently, Elliott *et al.* (2008) have proposed more efficient diving as an additional or alternative advantage of weight loss in this guillemot, although this could not apply to the kittiwake.

Prospecting females (i.e. individuals mainly three to four years old, which would breed in the next few years) also showed a drop in weight over the same period, although obviously this was not correlated with breeding. However, these individuals go through a similar but reduced ovarian cycle and hormone release as the breeding females, which are probably involved in bringing about this weight loss.

In contrast, prospecting males did not show the drop in weight at the time it occurred in other individuals. Unlike the prospecting females which were at a similar weight to breeding females, the prospecting males had consist-

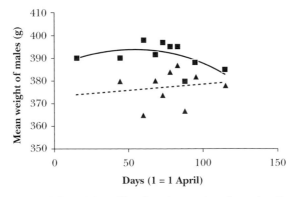

FIG. 7.3   *Comparison of the weights of breeding (squares) and non-breeding (triangles) male kittiwakes in relation to date. Note that non-breeding, prospecting males were consistently lower in weight than breeding males on the same dates. After Coulson (2010).*

ently lower body weights than breeding males throughout the breeding season (Figure 7.3). Perhaps this indicated the additional effort potential male recruits have to make to obtain and then retain nesting sites in the colony, something the female prospectors need not do to the same extent.

CHAPTER 8

Factors affecting individuals within a colony

TIME OF BREEDING WITHIN A COLONY

BREEDING performance within a kittiwake colony differs dramatically between early and late breeders. To many, this might appear to reflect the difference between the older, early breeding pairs and the younger, late breeding pairs, but this would be a simplistic view. Evidence accumulated during my studies suggests that other factors are also involved. This chapter specifically examines the breeding performance within the kittiwake colony at North Shields, north-east England, in relation to date of the start of egg laying by each pair.

TIME OF RETURN

The annual return of individual kittiwakes to the colony was spread over many weeks. The first birds to return to the study colony at North Shields often returned in January, while the birds which would breed late in that year did not arrive until early May, at which time some pairs in the colony already had built nests and laid eggs. As a result of this, there was a spread of 120 days during which time breeding individuals were returning to the colony. Further, non-breeding prospecting kittiwakes arrived even later, extending the arrival period to some 200 days.

The most important factor influencing the time of return was the age (breeding experience) of the individual birds, with the oldest usually returning first. Age for age, males and females returned at much the same time, with males a few days earlier, although this difference could be attributed to males reoccupying the sites first, which were then visited by the females.

An example of the prolonged period of arrival of kittiwakes that would breed later in that year is shown in Figure 8.1. Typically, the oldest birds returned first and were progressively followed by those with less experience, while the least experienced birds (those which were about to breed for the first time) arrived an average of eight weeks after the experienced breeders. The last of those arrived in the colony for the first time in early May. This drawn-out reoccupation of the colony had clear consequences. The older (experienced) birds had first choice of sites (and they often returned to the site they used in the previous year) and also mates. Young, inexperienced birds returned last and therefore had to use nest sites which remained unoccupied in the colony, either at sites where the adults which had used them previously had died, or on sites that had not previously been used for breeding.

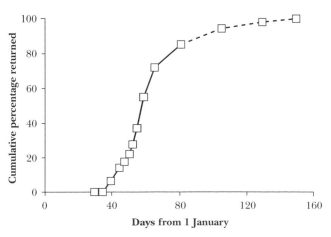

FIG. 8.1   *The cumulative date of first return of kittiwakes about to breed at the North Shields colony, 1960–1965. 95% of birds which had bred previously had returned by 31 March (day 90). The dashed line represents the arrival of birds about to breed for the first time, with a few still arriving after 1 May (day 121), by which time some of the oldest birds had started to lay.*

TIME OF LAYING

At the North Shields colony, laying started earlier and was spread over a longer period of time than in many other colonies in north-east England and south-east Scotland. As with the time of return, the date of laying differed markedly according to the breeding experience of the birds and there was clear evidence that the date of return influenced the time of laying.

Figure 8.2 shows the average breeding experience (including the current year) of females laying in each weekly period of the breeding season; the pattern for males was very similar. Typically, older females laid in the first half of the breeding season, in early or mid-May, but the first individuals to lay each year were never the oldest females, but usually middle-aged individuals.

First-time breeding females were not restricted to laying in the late part of the breeding season, although this is a commonly held view, and they showed the greatest variation in laying date than in any of the age classes. Females breeding for the first time laid throughout the entire span of the breeding season, but they also formed an increasing proportion of laying birds as the season progressed (Figure 8.3). By June, almost half those laying were first-time breeders.

The first-time breeding females that laid early in the season had often paired with older and experienced males, while the mates of those laying late in the season were invariably equally inexperienced. What determines the time of laying? In birds generally, males produce viable sperm long before females are capable of laying eggs. It is the female's development which limits the time for which eggs can be produced. Obviously, factors influencing this must occur sometime before laying and affect the rate of ovarian development and the date by which egg production can occur. The main factors

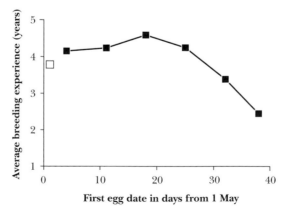

FIG. 8.2   *The average age (breeding experience, including the current year) of female kittiwakes in relation to the date that the first egg of the clutch was laid. Note that on average the birds were oldest in the middle of the breeding period. Based on 2,141 breeding attempts from 1960 to 1990. The open square indicates a small sample of seven individuals that laid on or before 1 May.*

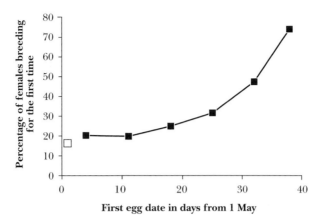

FIG. 8.3   *The percentage of first-time breeding female kittiwakes among those starting to lay in each seven-day time period. Based on 2,141 breeding attempts from 1961 to 1990. The open square indicates a small sample of seven individuals.*

involved are probably the time of return to the colony, because this is the start of courtship (influenced by the male), followed by the amount of stimulation received from neighbouring pairs. Prior to the return to the colony, the female's ovaries have started to develop, but once back in the colony and paired, the rate of development is increased. It is at this stage that I believe social factors play an important role. Because of their early return, older birds received more stimulation than younger ones and tended to be early layers. If this pattern is examined in detail, then it becomes obvious that there is indeed a close relationship between the time of return and the time of laying (Figure 8.4). The difference in the slopes of the two diagonal trend lines in this figure indicates that for every 3.5 days for which the time of return to the colony is delayed, the date of egg laying is one day later.

This explanation is compatible with the earlier return and start of laying at North Shields, compared to many other colonies. It also explains why in years when arctic conditions are severe in the spring, a delayed return to the colony is accompanied by later breeding.

In passing, Figure 8.4 reveals another interesting point. The two sloping lines meet at about day 176 (25 June), that is, when the date of return and breeding would coincide. Although such extrapolation is often dubious, it does suggest that kittiwakes could not possibly lay after 25 June in the North Shields situation. Allowing time for mating and nest building (about 14 days), this produces a date which is extremely close to 10 June, the average annual last laying date in the colony.

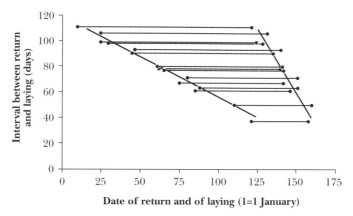

FIG. 8.4   *The relationships between the interval between returning and laying in individual female kittiwakes and the actual date of return (left-hand sloping line) and the date of laying (right-hand line). The difference in the slopes indicates that a change of about 3.5 days in the date of return changes the date of laying by one day. These regression lines are based on data for 256 individual females and a sample of the data for 17 of the individuals (reduced from 256 for clarity) is shown by the horizontal lines.*

CLUTCH SIZE

At North Shields (and elsewhere), clutch size varied markedly with the date upon which the first egg of the clutch was laid. In every year, from 1954 to 1990, the first clutch to be laid in the colony contained two eggs, and two eggs remained the most frequent clutch size laid in each time period until about 7 June, which is only three days before the normal cessation of laying there. Three-egg clutches only occurred in the first half of the laying season, and these were responsible for the trend (shown in Figure 8.5) in the average

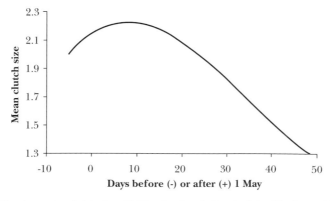

FIG. 8.5   *The change in clutch size of kittiwakes in relation to date of laying at North Shields, 1954–1990. The curve is the best fit to 2,664 clutches.*

clutch size with date at North Shields, which was repeated year after year. There was a small increase in clutch size early in the laying period, followed by an appreciable decline during the second half of the laying period, when the clutch size decreased from an average of about 2.2 eggs in the second week of May to 1.3 eggs per clutch in the few birds starting to lay in mid-June.

Although it might have been expected that the seasonal trend in clutch size would be relative to the date of the start of laying in the colony and just shifted in time, this was not so. This is evident in Figure 8.6, where the trend lines for clutch size against date for the earlier laying decade, 1961–70, are compared with the seven days' later onset of laying in the 1981–90 period. Apart from a slightly lower clutch size in the first half of May, the decline in clutch size almost coincided with the calendar date and differed by less than a day, rather than the seven days' delay which existed in the onset of laying.

This suggests that actual date, not relative date, determined the seasonal trend in the clutch size of the kittiwake. Support for this conclusion is illustrated by Figure 8.7, where the average clutch sizes in weekly periods in several other colonies are shown. All of these started laying later than at North Shields and it is evident that the trend lines sit close to (but very slightly higher than) that found at North Shields. The first birds to lay in these other colonies produced the largest average clutch size and in all cases this was followed by a progressive decline, matching the North Shields data for the same dates. As with the last laying females at North Shields, these also laid an appreciable proportion of one-egg clutches. Three-egg clutches were less frequent in all of these colonies than at North Shields.

This date effect also influences the average clutch size produced in any given year within a colony. Thus the average clutch size was lower at North Shields in the years when breeding began later, and the average clutch size at

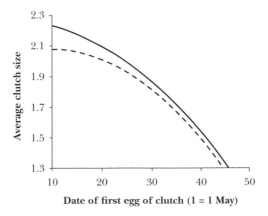

FIG. 8.6   *The trend by date in clutch size of kittiwakes at North Shields 1961–1970 (continuous line; based on 840 clutches) and 1981–90 (dashed line; based on 862 clutches). The data for the earlier laying in 1961–1970 have been truncated to start from 10 May to cover the comparable period in the 1981–1990 data when egg laying started a week later.*

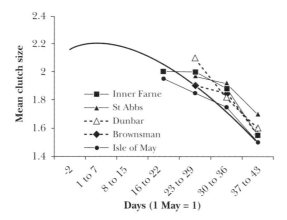

FIG. 8.7 *A comparison between the change in clutch size in kittiwakes in five other colonies in eastern Britain and the data from North Shields (continuous line). Note that the other colonies all bred later and showed a progressive decrease in clutch size which closely followed the similar decline at North Shields over the same dates. The mean clutch size was 2.03 at North Shields, and 1.86 at Inner Farne and Brownsman on the Farne Islands. In Scotland the means were 1.89 at Dunbar, 1.87 at St Abbs and 1.91 on the Isle of May.*

North Shields was higher than in later breeding colonies presented in Figure 8.7. Perhaps the best example of this is reported by Belopolskii (1957) who showed that in 1937, when breeding by kittiwakes was early on Kharlov Island in Arctic Russia, the clutch size averaged 2.33 eggs (n = 216), with 40% of the clutches having three eggs. This is the highest average clutch size recorded in a colony of kittiwakes. In the following year, breeding was late and the average clutch size was reduced to 1.53 eggs (n = 152), and not a single three-egg clutch was recorded.

It has been possible to explore the effect of date on clutch size using the run of data from successive years obtained for individual female kittiwakes at North Shields. Many of these birds produced the same clutch size in successive years and so contributed nothing towards testing the possibility that clutch size tended to be date-related. However in those females that did show variation between years, there was a clear tendency for those which increased their clutch size to have started laying earlier, while those which laid smaller clutches also laid later (Table 8.1). The more days by which females changed their date of laying, the greater was the chance of a change in clutch size. While the pattern clearly indicates that later laying by individuals was also associated with a tendency to lay a smaller clutch and early breeding with a larger clutch, there were also a few exceptions.

Table 8.2 shows examples of clutches laid by individual females in successive years and the tendency that a larger or smaller clutch size is closely linked to the date of laying. It would appear that individual females varied considerably in clutch size and in their response to a change in the laying date. For example, female 2044874 produced two eggs annually throughout 14 years,

TABLE 8.1 *The change in the clutch size of individual female kittiwakes between pairs of years in relation to the change in the date the clutch was started. Positive dates indicate earlier laying in the second year. The differences between the three categories are highly significant (P < 0.001). Data 1960 to 1975.*

Clutch size in consecutive years	N	Change in laying date (days)	Number laying earlier	Number laying later
Increased	48	+3.91 ± 0.72	38	10
No change	218	+0.62 ± 0.36	108	110
Decreased	50	-3.73 ± 0.69	12	38

despite the date of laying changing over a range of 26 days. Female 2020003 produced three-egg clutches in 12 of 13 breeding years, suggesting that some individuals were much more likely to lay larger or smaller clutches than others, an effect that was confirmed by a detailed statistical examination of the long data set.

These results, together with those for five other colonies shown in Figure 8.7, offer persuasive evidence that the trend in clutch size within kittiwake colonies is determined by the actual date of laying, and this was also evident in individuals. Since this pattern was repeated year after year, the females must be responding to an environmental clue which is expressed in a very similar way each year. Obviously, kittiwakes do not have a calendar, and the

TABLE 8.2 *Six examples of the date of laying of the first egg and clutch size produced each year by individual female kittiwakes in successive years. Note that, in general, a larger clutch size was associated with earlier laying and a smaller clutch with later laying by the individual, but the relationship was not perfect. Female 2044874 was consistent over 14 years in producing only two-egg clutches although the date of laying varied by up to 27 days, while 2020003 laid three-egg clutches in all but the last year of breeding, which was also the latest date on which she started laying. The date indicates the date when the first egg of the clutch was laid with 1 May = 1.*

EC11340		2020003		2044874		EC11418		EC11773		ED76629	
Date	Clutch	Date	Clutch	Date	Clutch	Date	Clutch	Date	Clutch	Date	Clutch
26	1	14	3	2	2	50	1	28	1	29	1
14	3	10	3	13	2	14	2	21	1	23	1
34	2	10	3	8	2	15	2	14	2	16	2
30	2	6	3	8	2	30	2	12	2	29	1
24	2	15	3	9	2	21	2	14	2		
29	2	15	3	2	2	21	2	17	2		
37	1	8	3	1	2	42	1	25	2		
22	2	7	3	4	2			7	2		
32	2	7	3	9	2			9	2		
32	2	8	3	5	2						
18	3	15	3	10	2						
32	2	12	3	28	2						
29	2	20	2	14	2						
				7	2						

most likely cue is probably linked to day length. But when the clutch sizes of kittiwakes are considered in more northern regions, it is evident that day length as such cannot account for the same pattern of change in clutch size in the Arctic, because the relationship of clutch size and date is shifted by two or three weeks, yet the average clutch size is not reduced in the Arctic. I believe that day length is an important factor in influencing clutch size in the kittiwake, but it is necessary to express this in a way other than by the actual length of the light period during the day. An additional but linked variable is needed to explain the cause of lower clutch sizes in later breeding birds within northern colonies. For example, the key factor could be the decreasing rate of change of daylight and this would also need to include an effective lapse time of 1–2 weeks to produce a universal explanation of the clutch size and its seasonal variation over the whole range of the kittiwake.

It is possible that with a better understanding of the environmental influences on clutch size and laying, it may be revealed to be the same mechanism that causes the cessation of egg laying in the colony, which again is an effect which does not seem to vary from year to year in the same colony, but the date is different further north. It is not known what environmental factors influence the few pairs that build nests each year, yet are inhibited from laying. These events occur in all colonies, and in most years, and the inhibiting factor is probably again related to day length. Unfortunately, kittiwakes are not good subjects for experimental investigations of such effects.

Egg size and date

The average volume of an egg laid by kittiwakes declined as the laying season progressed. This is illustrated in Figure 8.8 for clutches of two eggs, which

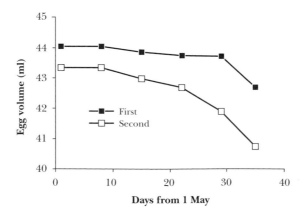

FIG. 8.8 *The volumes of the first and second eggs in clutches of two laid by kittiwakes in relation to the date of laying of the first egg. Note that the difference in volume between the two eggs tended to increase in those eggs laid late in the breeding season. Based on 2,664 clutches.*

formed over two-thirds of the clutches laid by kittiwakes during the North
Shields study period. The volume of the egg changed only marginally during
most of the breeding period, but then declined more rapidly in the late
breeding birds. On average, the second egg was smaller than the first
throughout the laying period, and the difference increased in the later laying
birds. Single-egg clutches showed a similar and slightly more pronounced
trend towards smaller eggs laid late in the season, and this also occurred with
the third egg in clutches of three. Thus in all clutch sizes, the size of the last
egg in the clutch declined to a greater extent as the laying season progressed.
It was as if the 'tap' controlling material being transferred into the last egg
was turned off earlier in later laying birds and, in extreme cases, it may have
resulted in some potential second eggs not reaching an adequate size for
laying and these were absorbed rather than being laid as very small eggs.
Presumably, such an effect would explain the marked increase of single-egg
clutches late in the season.

The shape index (breadth/length) of eggs also tended to decline with
date, and more so in the last egg of the clutch, indicating that these eggs
became relatively long and narrow. This effect was not as marked with respect
to date as it was within the position in the laying sequence. Possibly it was
produced by the oviduct starting to shrink from its maximum diameter
(perhaps with the onset of regression of the ovaries and related structures)
before the last egg was laid. A consequence of this would be that with the
same volume of content, the egg had to become longer.

BREEDING SUCCESS AND PRODUCTIVITY

The changes in clutch size, numbers of young hatched and chicks fledged in
relation to the date on which the eggs were laid are shown in Figure 8.9. The
figure does not show individual points (the lines are based on 2,664 points for
each variable), but the curves are the best-fits derived from individual points.
The pattern for clutch size has already been discussed. The curve (thin line)
for number of young hatched differed markedly in several respects from the
clutch size curve, indicating that hatching success was relatively low and
approached total failure in eggs laid in the second week of June, near the
date when egg laying ceased. The number of young fledged per nest (productivity) closely followed the curve for the number hatched, indicating a high
success rate from hatching to fledging. Productivity declined from over 1.4
young per nest from eggs laid in early May to zero in late clutches laid in
June, with the last laid clutches invariably failing during incubation, rather
than in the chick stage. This suggested once again that the poor breeding
performance of late laying pairs was not due to a food shortage for the young,
but to a failure occurring during incubation.

Figure 8.10 shows the percentage breeding success (number of young
fledged x 100 divided by the eggs laid) according to the date on which the

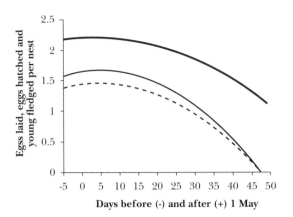

FIG. 8.9   *The pattern of change in clutch size (thick, continuous line), numbers of young hatched per nest (thin line) and numbers fledged per nest (dashed line) for each pair of kittiwakes in relation to the date of egg laying at North Shields. Based on 2,664 clutches from 1954–1990.*

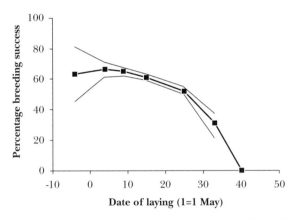

FIG. 8.10   *The percentage breeding success of kittiwakes nesting at North Shields 1954–1990 in relation to the date of egg laying (first egg). The thin lines indicate one standard error from each point. Note that the last 17 pairs to lay failed completely.*

eggs were laid. The figure demonstrates a dramatic decline from about mid-May, when it was just over 60%, to zero in the last clutches to be laid in June. Those laying in the first half of the breeding season had a larger clutch size, larger eggs and a high breeding success, and so were highly productive and accounted for most of the productivity within the colony.

The later breeding pairs in the second half of the laying period (after 19 May) produced less than half the number of fledged young per pair compared with those laying in the first half of the breeding season. In part, the poor breeding performance of late breeding pairs can be attributed to the greater proportion of young, first-time breeders late in the season, but consideration of the average age of individuals in Figures 8.2 and 8.3 raises a doubt as to whether this is the only reason for the poor breeding performance of the late breeders. The decline in productivity is too large to be explained by the change in age structure alone.

This conclusion is confirmed by the examination of the breeding performances of first-time breeding females and experienced breeding females in relation to the date of egg laying. Figure 8.11 shows the clutch size in relation to the date the first egg of the clutch was laid in first-time breeding females and also for a comparably large sample of birds which were breeding for more than the fourth time. Clearly, date for date, the older and more experienced birds laid larger clutches, but later breeding individuals of both groups showed a decline in clutch size, confirming that the smaller clutches at the end of the breeding season were not solely the effect of the age of individuals. The breeding success of these two age classes showed similar trends (Figure 8.12), with differences in success between the two categories on the same date, but both showed a progressive decline in success as the breeding season advanced. In both categories, the major (and significant) drop in breeding success occurred among the birds laying at the end of the season.

There are other variables which cause relatively small changes in clutch size and breeding success, including whether birds are at the centre or edge of the colony, have a male partner of similar or different age and, in older birds, whether they had taken the same mate as in the previous year (see later). Yet even when these were taken into account, the downward trends in the breeding performance among later breeders still remained large.

So what are the main causes of the downward trend? The indications from the series of measures of breeding performance considered all indicate a deterioration as the date of the laying period progressed. The most likely explanation is that later breeders tend to be birds in poorer condition as well as young, less experienced individuals, and that both these factors tend to

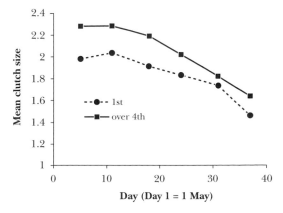

FIG. 8.11 *The change in average clutch size with date of first-time breeding female kittiwakes (sample of 822 breeding attempts) and those breeding for more than the fourth time (873 breeding attempts).*

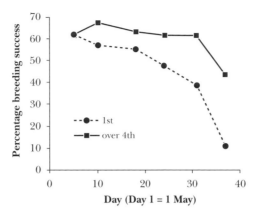

FIG. 8.12   *The change in breeding success with date of laying in first-time breeding female kitti-wakes (sample of 822 breeding attempts) and females breeding for more than the fourth time (873 breeding attempts). Note that the grouping of data has been changed from that in Figure 8.10 to give larger samples for the extreme points.*

induce a smaller clutch size, lower breeding success and reduced production of young reaching fledging age. It should be noted that not all young birds bred late, and those which bred earlier were more productive, so quality of the individual appears to be affecting some young birds.

In every aspect of the breeding biology of kittiwakes which has been examined, there is a deterioration in the performance with date. The last birds to lay not only produce fewer and smaller eggs, but their breeding success is appreciably reduced and their contribution to the productivity of the colony approaches zero. In several bird species, similar deterioration in the breeding performance has been reported in the last to breed, but in many of these cases, the poor breeding performance of the late breeding pairs has (rightly) been attributed to food shortage caused by the birds hatching their eggs too late to exploit a brief peak of food available to the earlier nesters. In the case of the Great Tit *Parus major,* when the eggs of late breeders hatched they had missed the peak of abundance of the rapidly growing Winter Moth caterpillars, when most of the insects had already left the trees and pupated in the ground, thus producing a period of food shortage (Gosler 1993, van Noordwijk *et al.* 1995). However, this explanation does not apply to the kittiwake, whose main food sources survive and are available for much longer periods. Fish stocks are much more stable than those of caterpillars because fish produce overlapping generations extending for several years, tending to produce a greater degree of stability in numbers. In fact, the food available for kittiwakes usually increases appreciably during the summer, as the fish eggs spawned early in the year have hatched, and the larvae (feeding on spring plankton blooms) grow rapidly into fish of a size which is readily exploited by kittiwakes and other seabirds before midsummer.

The ineffective breeding by late layers suggests that the alternative strategy of missing a breeding year would not necessarily be an appreciable disadvantage to these individuals, and this provides a plausible explanation for why some adult kittiwakes do miss breeding, particularly when they are young.

An obvious question remains to be answered: what happens in other kittiwake colonies where the onset of laying is later and breeding is even more synchronised than at North Shields? General observations made in other colonies strongly suggest that the pattern of decreased success with date is similar to that found at North Shields, but it is more truncated over time, and the major contribution to the total production of young by the earlier breeds is maintained.

EFFECT OF AGE AND POSITION IN THE COLONY

Early in the study, it became evident that two factors had an appreciable impact on the breeding performance of individual pairs of kittiwakes, namely the age of the adults and the position the pair within the colony. In this analysis, I did not know the actual age of many of the breeding birds because they were unmarked when they first bred in the colony and there is no method of aging a fully adult bird, but I did know that when they first bred in the colony, they were breeding for the first time. As I collected more records of individuals which had been ringed as chicks, it was clear that most started breeding when they were three to five years old. To fully use the information collected, I have therefore used breeding experience as a surrogate of actual age, so most birds were about four years older than years of breeding experience. In what follows, I use 'age' for convenience to indicate years of breeding experience. To obtain adequate sample sizes of the older individuals, several of the older age classes were grouped together. This did not create a problem since most of the differences in breeding performance proved to be among the younger birds and this allowed me to look for differences in the performance of females of comparable age breeding in different parts of the colony.

The identification of differences in the performance of kittiwakes breeding at the centre and at the edge of a colony (published in the scientific journal *Nature*, Coulson 1968) was the first time that such an effect had been identified in a colonial bird species. Since that publication, I have accumulated many more data and these have been reanalysed in this book. Although I believe this effect (between the central areas and edge) is actually a gradient and relative to the proximity of the edge of the colony (see later), it has been convenient to restrict the classification of sites in the colony to two groups to ensure that the analysis is based on large samples of data for birds of different breeding experience. I call these two groups 'centre' and 'edge'. To avoid a personal bias in deciding which sites formed the centre and edge, I defined the centre of the colony at North Shields as that which was occupied when the colony contained only 45 nests. Soon, the colony doubled in size and expanded at the horizontal edges and to a lesser extent at the top and bottom of the colony. Over the subsequent years, this produced approximately equal numbers of records of breeding at the centre and edge. It is important to note that centre and edge sites were on similar sized ledges, each with the same physical characteristics, and these only differed in their position within the colony. Further, nest predation was negligible throughout the colony and did not play any part in the differences found.

Only in six of over two thousand breeding occasions did an individual male or female cross the boundary defined as edge and centre. An edge bird was an edge bird for life and therefore was permanently influenced by the decision as to where to breed for the first time. This meant that it was not possible to examine the performance of the same individual when switching between the edge and centre, or vice versa. The breeding results detailed here are based on females. Males showed very similar results as for females in relation to their age and position in the colony and it is unnecessary to repeat these in detail, while adding the data for the males to that of the females would involve using most of the breeding data twice. The results are based on 2,323 breeding attempts. With respect to the time of return to the colony, I have presented information for both sexes since it is highly likely that the pairs keep together during the winter and each makes an independent decision as to when to return.

In the sections below, the breeding performance of kittiwakes has been analysed both by age and by position in the colony, illustrating that both had an appreciable effect on breeding. I have also restricted the data for older females to those who, in that year, were breeding for the first time with the partner, to make the data strictly comparable with first-time breeding females, all of whom were also breeding with that male for the first time. The effect and benefit of retaining the partner from the previous year is considered later.

1. Effect of age and sex on date of return to the colony (Figure 8.13)

Both males and females arrive late when about to breed for the first time and on the same average date whether they were about to nest at the centre or the edge (Figure 8.13). Birds returning to breed for the second time arrived about ten days earlier than in the previous year, but again there was no meaningful difference in the dates between central and edge nesting birds of either sex. The more experienced birds nesting at the centre and the edge showed a marked difference in their return dates, with those breeding for the third to fifth times arriving back an average of 15 days earlier than the comparable edge birds. Note that the distinctly earlier return date of birds of the same age nesting in the centre and at the edge of the colony was not acquired until the birds had already bred in at least two previous years.

This delay in the appearance of a difference was also found in other aspects of the breeding biology of centre and edge birds (see below) and is discussed later.

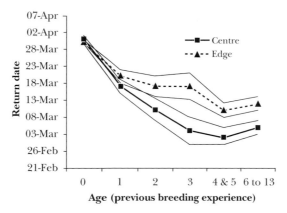

FIG. 8.13   *The average day of return to the colony of male and female kittiwakes breeding in the centre and at the edge of the colony in relation to their previous breeding experience (age). The thin lines show one standard error from the means. There were no appreciable or consistent differences between the sexes.*

2. Date of laying (Figure 8.14)

The most extensive data are for females breeding for the first time, and are based on over a thousand individuals. These females laid late, but at the same average date in the centre and at the edge of the colony. Females breeding for the second time laid earlier but there was still no difference in the mean values for the centre and edge. In older birds, those nesting in the centre lay consistently earlier than those at the edge, although the effect may be lost in the oldest females. Apart from the effect of position in the colony, birds

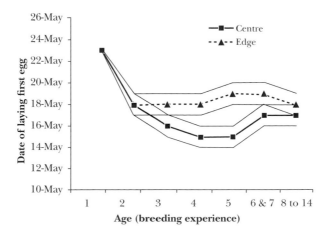

FIG. 8.14   *The mean date of laying the first egg of a clutch in female kittiwakes of different ages in relation to position at the centre and edge of the colony. Data for females which have a new mate. The thin lines show one standard error from the means.*

breeding for more than the second time show little evidence of an advancement of the average laying date.

3. Clutch size (Figure 8.15)

The mean clutch size laid by females increased gradually with age, but did not differ between the centre and edge for those females breeding for the first, second or even the third times. Thereafter, the average clutch size was consistently and significantly higher in the four oldest age classes of centre females, but the difference was only about 0.1 of an egg, which represent the centre females increasing egg production by 5% over the edge females.

4. Breeding success (Figure 8.16)

The percentage breeding success from egg to fledging was relatively low but similar in first-time breeding females both in the centre and at the edge of the colony. It improved when the females were breeding for the second time but still did not differ between the centre and edge. Thereafter, the older birds at the centre maintained a high breeding success of over 63% while those at the edge remained consistently lower, averaging 56% (Figure 8.16). This difference represents about an 11% greater breeding success in the older birds nesting at the centre of the colony. It should be emphasised that this difference was not caused by predation of the eggs or young.

5. Productivity (Figure 8.17)

Since clutch size varied relatively little between the centre and edge of the colony, the productivity (young fledged per pair) tended to follow the

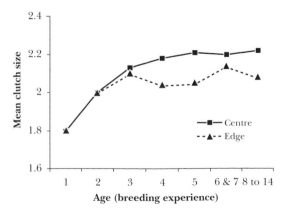

FIG. 8.15   *The mean clutch size laid by female kittiwakes of different ages in the centre and edge of the colony. Note that the average clutch was two eggs or larger in all age classes except for those females breeding for the first time. Data are for females that had a new mate.*

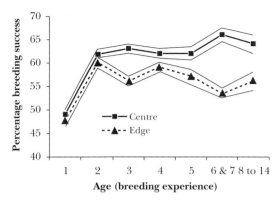

FIG. 8.16 *The percentage breeding success of female kittiwakes of different ages in relation to whether they breed in the centre or edge of the colony. Data only for females that have a new mate. The thin lines indicate one standard error from the mean values.*

changes in breeding success. Young females breeding for the first two occasions appreciably increased their productivity, but did not show a difference according to position in the colony. Each female breeding at the edge produced about 1.20 young to fledging during the second and subsequent breeding attempts, while the comparable females at the centre moved progressively to an even higher level of productivity with an average of 1.35 young fledged per pair, that is, a 13% higher success. The higher productivity at the centre is produced by a combination of a slightly higher clutch size followed by an appreciably greater breeding success.

In every year over a 31-year study period, kittiwakes breeding at the centre of the colony had a greater productivity than those near the edge, and Table 8.3 shows the average differences between the two areas over this period of

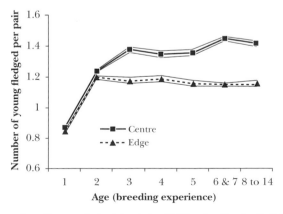

FIG. 8.17 *The number of young fledged per pair of kittiwakes (productivity) at the centre and edge of the colony in relation to female age. Data for females that had taken a new mate. The thin lines indicate one standard error from the means.*

TABLE 8.3 *The annual difference in the average breeding performance of all kittiwakes irrespective of age, breeding in the centre and at the edge of the North Shields colony, 1960 to 1990.*

	N	Return	Laying	Clutch	Breeding success	Fledged (productivity)
Centre	1140	16 Feb	17.3 May	2.10	64%	1.35
Edge	1070	18 March	21.9 May	1.98	61%	1.21
Difference		30 days	4.6 days	0.12	3%	0.14
Percentage increase at the centre				6%	5%	12%

time. What are the causes of these differences? One possibility is that they are the result of age differences in females nesting in the centre and at the edge of the colony. I have determined the age of females nesting in the two areas in three years, each ten years apart (Table 8.4). In all three years, there were more young females and fewer old females at the edge than at the centre, but when the effects of these age differences are evaluated (Table 8.5), they contributed little towards explaining the much larger differences between the centre and the edge. At best, age composition explained 21% of the greater productivity at the centre of the colony. Clearly some other effects were causing most of the difference between centre and edge.

These figures contribute substantially to understanding the benefits obtained by nesting in the central part of the colony. The benefits are not immediate, as might be expected had the differences been an effect of better quality individuals obtaining sites at the centre of the colony, with less good individuals restricted to the edge areas. Of particular importance is the fact that the differences between centre and edge females did not become appreciable until they had bred for two occasions, and often were not fully developed until they bred a third time. In other words, the difference appeared to be progressively acquired in response to differences in condi-

TABLE 8.4 *The percentage distribution of the age (breeding experience) of kittiwakes breeding at North Shields in 1965, 1975 and 1985 at the centre and at the edge of the colony.*

	First	Second	3rd to 4th	5th to 7th	8th to 10th	Over 10th	Average breeding experience
1965							
Centre	24%	17%	23%	20%	4%	12%	4.51
Edge	32%	10%	29%	23%	5%	1%	3.93
1975							
Centre	19%	16%	18%	23%	15%	9%	5.04
Edge	22%	19%	22%	19%	9%	9%	4.49
1985							
Centre	41%	12%	23%	16%	3%	5%	3.34
Edge	50%	17%	20%	8%	4%	1%	2.48

TABLE 8.5 *The difference in breeding performance at the centre and edge of the colony which can be attributed to the age compositions shown in Table 8.4. The values given are the centre means minus the edge mean values. All values were earlier or higher at the centre of the colony.*

	Return (days)	Clutch size	Laying date (days)	Productivity (young per pair)
1965	1.6	0.010	0.6	0.06
1975	1.1	0.010	0.9	0.02
1985	0.9	0.008	0.5	0.02
Average	1.2 days	0.009	0.7 day	0.03
Observed difference	30 days	0.120	4.6 days	0.14
Percentage of observed difference explained by age composition	4%	8%	15%	21%

tions at the centre and edge, and was not attributable to inherent differences in individuals evident at the time of recruitment as breeding birds.

There was no effect of site position on first-time breeding females, presumably because these individuals arrived late at the site. This suggested the influence of a social effect operating during the pre-laying period. When the young birds arrived, most birds in the colony had already completed most courtship and were nest building and laying. As a result, the young birds would have received little social stimulation from the courtship of their neighbouring birds. The minimal effect of position on the second breeding occasion is similarly explicable.

THE BENEFIT OF RETAINING THE SAME MATE IN RELATION TO AGE

There was an advantage in retaining the mate from the previous breeding season and the benefits were the same for both centre and edge nesting birds. The retention of the same mate had several effects; it advanced the date of laying and increased both the breeding success and productivity.

First-time breeders usually had visited the colony in the latter part of the previous breeding season, and some even formed temporary pairs, but they invariably paired with a different mate when they eventually bred for the first time in the subsequent year. As a result, all first-time breeding females bred with a new partner. When making comparisons of the breeding performance of older kittiwakes, they were made only on females which had changed their mate from the previous year.

There were two reasons for which established breeding kittiwakes took a new mate. In some instances, the partner failed to return to the colony and presumably had died since the last breeding season. In other cases, both members of the pair from the previous year returned to the colony in the spring and bred, but both took different partners, a situation which I have called 'divorce'. While changing mate between years had an adverse effect on breeding performance, the effect was the same whether caused by divorce or the death of the previous partner.

Females breeding for the second time improved their breeding performance from the previous year, but there was no difference in the time of laying, clutch size, breeding success or productivity according to whether they retained the same mate from the previous year or bred with a new male.

Older females that retained their mate from the previous year laid an average of two days earlier than those with a new partner, but the clutch size was not meaningfully greater until they had at least five years of breeding experience (Figure 8.18), but even then, the average increase was small and less than 4%.

Females breeding for the second time obtained no advantage at any stage of breeding by retaining their partners from the previous year. In contrast, those breeding for at least the third time which had retained their mate from the previous year had nearly 20% greater breeding success and productivity than those which took a new partner (Figures 8.19 and 8.20). It is evident that retaining the mate from the previous year appreciably enhanced the breeding success and productivity of older females, but not of young females. Results suggest that familiarity with the mate was not the only factor important in producing greater breeding success and productivity, as the effects were only evident in the older females that had retained their mates. Retaining the same mate for three or more breeding seasons produced no further advantages.

How is an advantage achieved by retaining the partner from the previous year? One of the factors linked with divorce is breeding failure in the previous year, and breeding failure is, in some cases at least, associated with incompatible behaviour between the members of the pair. As a result, it can be argued

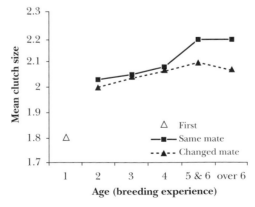

FIG. 8.18   *The mean clutch size laid by female kittiwakes according to breeding experience and whether they have taken a new mate or retained the mate from the previous year. The unfilled triangle shows the data for first-time breeding females, all with new mates.*

that there is greater chance of obtaining a compatible partner by taking a new mate rather than retaining the partner of the previous year who was involved with breeding failure. Because kittiwakes can identify their mates from previous years, the retention of a mate from the previous year may reduce the time needed to be spent in the necessary preliminary behaviour in first establishing a bond between the two individuals. It should be remembered that pairs that were successful in the previous year were more likely to stay together. The shift system during incubation varied between pairs and some-

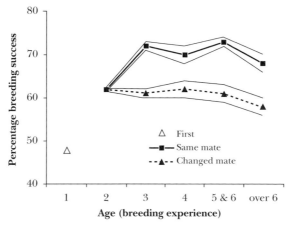

FIG. 8.19 *The percentage breeding success in relation to breeding experience of female kitti-wakes which changed mate or retained the partner from the previous year. The unfilled triangle shows the data for first-time breeding females.*

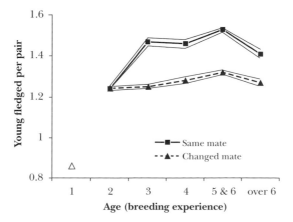

FIG. 8.20 *The number of young fledged (productivity) in kittiwake pairs that retained the mates from the previous year (squares and thick line) or changed mates (triangles and dotted line). The thin lines show one standard error of the means. Significantly greater productivity occurred in females when they bred for the third or more times and retained the same mate. The unfilled triangle shows data for first-time breeding females, all with new mates.*

times it broke down, presumably due to incompatibility. So there is an advantage in retaining a successful partner, and taking a new mate is a gamble and does not exclude incompatibility occurring later in the breeding processes.

The advantageous effects gained by retaining the mate from the previous year and nesting at the centre are additive. Thus females nesting at the centre and retaining the mate from the previous year bred earlier, laid a larger clutch and were more successful breeders than any other category, while females nesting at the edge and with a new partner were the least successful. An example of this additive effect for the date of laying is shown in Figure 8.21, with all of the centre-change partnerships and edge-retain partnership points intermediate in date compared with the extremes of the centre-retain and edge-change categories.

There is an important similarity between the effects of retaining a mate and the differences in breeding at the centre and edge in that the effects were not evident until the females were breeding for the third time. This suggests that the mechanisms which brought about the differences in breeding performance were likely to be the same, yet were not specifically produced by retaining the mate or the position in the colony, but rather that they were both influenced by the length of the period between forming the pair and laying, when courtship is at a maximum. In other words, the effects are probably determined by the date of return and establishment of the pair. This leads back to the importance of nesting density in the biology of the kittiwake.

There is convincing evidence that the density of other kittiwakes in the colony is a major influence in facilitating breeding and influencing the breeding success of individual pairs. The use of centre and edge is simply a surrogate for high and low density. Typically, the density of breeding kitti-

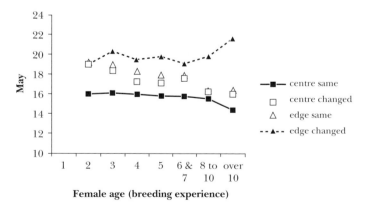

FIG. 8.21 *A comparison of the date of laying in relation to breeding experience of four groups of breeding female kittiwakes and whether they nested at the centre or edge of the colony and with the same or a different male partner from the previous year. Those keeping the same partner and nesting in the centre of the colony bred consistently earlier while those at the edge and with a new partner bred consistently late.*

wakes is higher in central parts of the colony than near the edge. This difference in density is enhanced by the fact that near the edge of the colony there are fewer or even no other breeding kittiwakes in one or more directions, and so the potential stimulation from other birds is less near the edge than at central sites.

The proportions of kittiwakes that changed mates each year, the number which nested in the centre and edge of the colony and the age distribution of breeding adults changed relatively little between years. As a result, these effects contributed only marginally to between-year differences in the time of return, time of breeding and breeding success within the colony as a whole. On the other hand, the effects of position reveal extremely important information about how the breeding performance of individuals is affected and determined.

The delay in the appearance of advantages of central sites compared with edge sites until a female has bred on two or more occasions is informative. It suggests that the difference in performance between central and edge individuals is not primarily an inherited characteristic, although there is much greater competition to obtain central rather than edge sites and this may introduce a partial separation of individuals with different drives to establish ownership of a central site. It seems more likely that the progressive development of differences in performance at the centre is the effect of stimulation from neighbours, and this increases in older individuals as they return earlier to the colony and so spend more time in social pre-breeding activities.

The earlier return of centre birds is interesting because it raises the question as to how this effect is produced. Is it an effect carried over from the previous breeding season, with birds tending to return to the same nesting sites, or does it directly involve a difference in the physiology or 'quality' of the birds? This question is discussed in more detail later.

SURVIVAL AND LIFETIME REPRODUCTIVE SUCCESS AT THE CENTRE AND EDGE OF THE COLONY

In animal species that can breed on several occasions, the breeding success in any given year is relatively unimportant when considering aspects which could be selected. Natural selection operates on the number of offspring produced in an individual's lifetime (or more correctly, the number of their young that survive to breed). In short-lived animals, the lifetime breeding success can be determined relatively rapidly with a few years of investigation, but in long-lived animals such as seabirds, it can take many years to collect representative data which cover the range of life spans of individuals. In the case of the kittiwake, some individuals in my study colony bred for up to 20 years before dying. Accordingly, a representative sample of lifetime productivity has to consist of samples from the study where all of the individuals, including the longest lived, have reached the end of their lives. In this investigation, all of the marked individuals which started to breed in years from 1954 to 1974 had died by the end of the study. Birds starting to breed in later years have been excluded, because they would bias the estimates.

In this study, I have not been able to record the number of young which survived to breed, because many of the young moved to breed in other colonies far from where they were reared and many were not found. However, I have made the assumption that the number which fledged (and at this time they ceased to be dependent on their parents in any way) is representative of the proportion which eventually survived to breed. This assumption is probably true, although the weight achieved at fledging may be an additional, but so far unknown, factor influencing post-fledging survival rates.

Survival rates of adults according to sex differed in this study and also differed between birds nesting in the centre and at the edge of the colony, and so both factors have been considered.

SURVIVAL RATES AND LIFE SPANS
OF FEMALES AND MALES

The number of years for which females and males survived after breeding for the first time in the centre or at the edge of the colony is shown in Figures 8.22 and 8.23, and separate survival curves have been drawn for birds breeding in the centre and edge. The percentage surviving each year has been plotted on a logarithmic scale because by so doing, the slope of the line indicates the mortality rates; the steeper the slope the higher the mortality rate, and changes in survival/mortality rates are visually evident at a glance.

The straight lines fitted to the data points in Figures 8.22 and 8.23 show that the annual survival rates in both sexes breeding at both centre and edge of the colony remained constant in relation to age (breeding experience) for

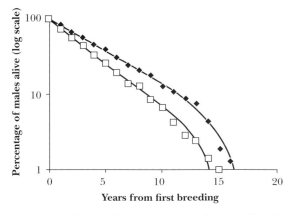

FIG. 8.22   *The percentage of adult males surviving from the time of first breeding at the centre (diamonds) and edge (squares) of the colony, based on 209 edge males and 158 centre males.*

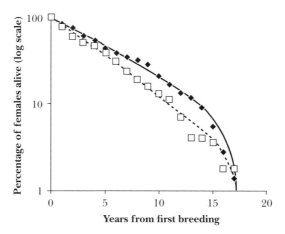

FIG. 8.23   *The percentage of females surviving from the time of first breeding at the centre (diamonds and thick line) and edge of the colony (squares and dashed line), based on 169 edge females and 145 centre females.*

at least the first 14 breeding occasions, but the differences in the slopes show that the rates differed between the sexes and by position in the colony. However, after about 14 breeding years, there is a suggestion that the survival rates decreased in the oldest birds in all categories, but by this time it affected only the longest lived 3–4% of the breeding birds and so this change in survival rate is based on relatively few individuals and is not meaningful.

The survival rates and expectations of breeding life of males and females nesting at the edge and centre of the colony are presented in Table 8.6, confirming the lower annual survival rates of males than females and between those breeding at the edge and centre. When these survival values are converted into the number of years for which the average individual in each category survives (and so is able to breed), the magnitudes of the differences between kittiwakes nesting at the centre and edge of the colony are more easily appreciated, with females at the centre of the colony living over a year longer than those at the edge, which increases their breeding life-span by over a quarter. The average difference between centre and edge males was even greater, with centre males surviving and breeding for a third longer.

As a result, not only do centre birds produce more young each year, but they also breed for more years. Thus, females nesting at the centre fledge 53% more young in their lifetime than those nesting at the edge and for males the difference is even greater, 63%. These are large differences and indicate the considerable advantage for birds obtaining a central site.

It should be noted that I found no evidence that young reared on central sites were more likely to return and breed in the centre of the colony. In fact, the data indicated they were slightly less likely to nest there, again suggesting that the quality of central birds was not primarily inherited, but something that was acquired either by the positions in which they nested or by the ability

TABLE 8.6 *A comparison of the annual survival rates and expectations of life from the time of first breeding of adult male and female kittiwakes nesting at the centre or at the edge of the North Shields colony.*

	Female annual survival rates	Female expectation of life (years)	Male annual survival rates	Male expectation of life (years)	Young fledged per year	Young per lifetime of female	Young per lifetime of male
Centre	84.3% ± 1.2%	5.83	81.1% 1.4%	±4.79	1.18	6.9	5.7
Edge	80.5% ± 1.4%	4.63	75.5% 1.5%	±3.58	0.98	4.5	3.5
Percentage centre greater than edge		25.9%		33.8%		53%	63%

of the individuals to develop efficient feeding methods and feed successfully during their immature years.

LIFETIME PRODUCTION OF YOUNG IN RELATION TO BREEDING EXPERIENCE

The number of young fledged during the life span of individual females which had bred at the edge or centre of the colony is shown in Figure 8.24. Many of the points at the lower left corner of the plot represent several individuals. The number of breeding years is the main factor determining the number of young fledged during the adult life span, and this explained 80% of the variation in female lifetime productivity. Only 8% of breeding females which laid at least one clutch of eggs failed to fledge young in their lifetime.

In addition to those birds which survived to breed, there were many individuals which fledged but did not live long enough to breed, and these plus the unproductive adults formed about 60% of all females that fledged, leaving 40% of the females which fledged producing at least one fledged chick. In contrast to the majority of females that produced only a few young in their lifetime, the most productive female fledged 28 young in 17 breeding seasons (Figure 8.24). The results for males were very similar to those for the females, but with slightly lower trends since males, on average, did not live as long as females.

Figure 8.24 shows the trend lines of different slopes for the number of young fledged in the lifetime of females in relation to the number of years each female lived, and these differed for those nesting at the centre and edge of the colony. In both cases, the main trend was linear, but this concealed differences in those breeding only once or twice.

This is confirmed in Figure 8.25, which shows that the annual productivity of females that survived to breed only once or twice was lower than that for the first two years of breeding in females which survived longer.

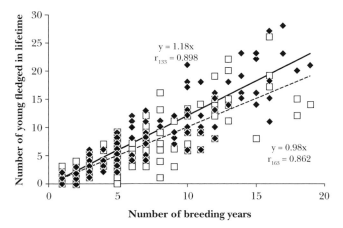

FIG. 8.24 *The number of young fledged in the lifetime of female kittiwakes. Eighty percent of the variation in number of young produced is explained by the number of years as a breeding adult. The black diamonds and thick line are individual values and the trend line for females breeding at the centre of the colony, and the squares and dashed line are for females breeding at the edge. Several of the points in the bottom left-hand corner of the graph each represent several individuals.*

There are two possible interpretations of this difference. One is that the females which died after their first breeding attempt were genetically poorer individuals. The second possibility is that the females that died in the year after breeding for the first time were in the early stages of an infection or

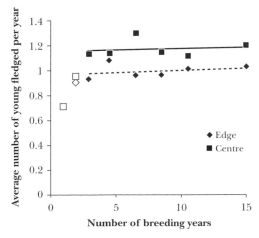

FIG. 8.25 *The average number of young produced each year by females which survived and bred for different numbers of years. Females which bred for only one or two years (open symbols) had lower productivity than older birds, but after the second breeding attempt there was little difference between the annual productivity and the number of breeding years. The trend lines for centre (solid line) and edge birds (dashed line) are essentially horizontal.*

disease that would kill them in the next year, and this hindered their breeding performance. These are not exclusive possibilities. The second of these is considered further in the next section.

Breeding performance in the year before death of one or both partners

Table 8.7 presents data for kittiwake pairs of all breeding experiences after the first occasion, and considers their last breeding occasion before they died. It is evident that, on average, females of all ages bred less successfully in the year prior to their death and that this breeding success was reduced even further in those cases where both members of the pair died before the next breeding season.

Confirmation that reduced breeding success in kittiwakes prior to death was evident in birds of all ages, and is shown in Table 8.8. This is an important

TABLE 8.7 *The last breeding performance of kittiwakes which were breeding for the second or more times and when one or both members died before the next breeding season. These are compared with breeding performance of comparably aged birds which survived to the next breeding season. In all cases, the last breeding performance was lowest in cases where both members of the pair died after breeding. Significance denotes meaningful difference within the column. There was no significance in the values according to which sex died.*

	N	Laying	Clutch	Young fledged	Percentage total failed	Breeding success
Both birds survived	451	20 May	2.12	1.42	15%	67%
Male died	222	19 May	2.06	1.23	23%	60%
Female died	204	20 May	2.01	1.24	22%	62%
Both died	135	18 May	1.99	1.01	34%	51%
Significance		no	yes	yes	yes	yes

TABLE 8.8 *The breeding performance of female kittiwakes according to their breeding experience and whether they died or survived to the next breeding season. In all age classes, those females which died had an appreciably poorer breeding performance.*

Female's breeding experience	N	Young fledged per pair	Young fledged per pair	Percentage decrease prior to dying	Percentage total failure	Percentage total failure	Percentage increase of failure prior to dying
		Survived	Died		Survived	Died	
1	132	1.17	0.68	42	25	51	104
2	66	1.37	1.07	22	17	30	76
3–5	71	1.39	1.11	20	15	33	120
Over 5	98	1.40	1.26	10	16	31	94

TABLE 8.9 *The breeding performance of the same female kittiwakes in successive years, the second year being the last breeding occasion before they died. After Coulson & Fairweather (2001).*

Breeding experience	N	Penultimate productivity	Ultimate productivity	Penultimate total breeding failure	Ultimate total breeding failure
1 to 2	89	1.10	1.12	31%	28%
2 to 3	54	1.36	1.20	20%	31%
3 to 4	45	1.48	1.05	15%	28%
4 to 5	32	1.71	1.50	11%	18%

result, because it indicates that the reduced breeding performance was unlikely to be caused by senescence, since that would have been indicated by the reduced breeding performance being restricted to the older birds. It is most unlikely that senescence occurred in kittiwakes breeding for the first time, and the adverse effect is reasonably attributed to the condition of the individuals in the weeks or (in most cases) several months before they died. This effect suggests that many kittiwakes die from a progressive deterioration over several months, rather than from a sudden traumatic event such as predation. The effect occurred in individual females, where their breeding performance in their last breeding occasion is compared with that a year earlier (Table 8.9).

The breeding performance of females that died in the year following their second breeding attempt was comparable with that in the female's first year of breeding, but overall, females breeding for the second time are much more successful than in their first breeding attempt, so this result indicates that the females breeding for the second and last time failed to show the improvement characteristic of their age. In older females, the last breeding attempt was consistently poorer than the attempt in the previous year. Clearly, many kittiwakes have a poorer breeding performance in the weeks or months before they die. It should be noted that none of these individuals died before their young fledged and, in many cases, the individuals were still alive weeks after fledging their chicks and presumably died in the winter period, when they were away from the colony and in the pelagic areas of the Atlantic Ocean. Death for these birds was not a sudden event; they deteriorated over weeks and months prior to their death, which affected their condition and last breeding attempt.

RETENTION OF NESTING SITES BETWEEN YEARS

I have left until last the consideration of the proportion of birds which retain their nest site between years, because it is the only variable considered that did not influence the breeding performance of individuals. No evidence could be found that retaining or changing site influenced the timing and success of breeding, other than those associated with age, and retaining or changing mates from the previous year.

I have already commented that breeding kittiwakes very rarely move between the centre and edge of the colony. In fact, the great majority of kittiwakes use the same nest site and even if they change sites, this usually involves only movement of a few metres. For example, although kittiwakes nested on all four sides of the warehouse at North Shields (see Figure 8.29 on page 171), extremely few moved from one side to another, and those that did usually moved just around the corner. In 35 years at North Shields, there was only one case of a bird moving to the opposite side of the building to breed.

The window ledges of the warehouse at North Shields were equally spaced and each was readily identifiable. As a result, it has been easy to use window spacing to quantify how far kittiwakes moved their nest sites between successive years. The distance between window sills was 1.4m, and the extent of change between the current and previous nesting sites has been recorded as the sum of sills, either horizontally, vertically or both. No appreciable or consistent effects of colony position (centre and edge) were found, and so the evaluation of site retention has been considered for the colony as a whole. The nest site used each year was recorded as that in which the eggs were laid, but the actual selection of the site took place soon after the annual return to the colony and often weeks or even months before courtship feeding, copulation, nest building and laying occurred. I have expressed breeding experience as the number of years of breeding before the selection was made in the current breeding season.

Overall, 22% of males and 34% of females which had bred and then returned to the colony the following year changed their nesting sites (Figure 8.26, Table 8.10), but the behaviour also varied markedly with the age of the individuals.

The proportion moving site decreased from 44% in females which had

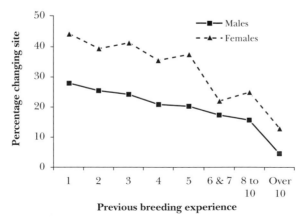

FIG. 8.26   *The percentage of breeding male and female kittiwakes at North Shields that changed their nest site in the following year in relation to their previous breeding experience.*

TABLE 8.10 *The percentage of male and female kittiwakes at North Shields that changed their nesting site in successive years, and the average distance moved by those which changed site, all in relation to their previous breeding experience. The average movement is the number of sills between the old and new site.*

Previous breeding experience	Males			Females		
	Sample size	Percentage moving	Average movement (sills)	Sample size	Percentage moving	Average movement (sills)
1	393	28.0	2.58	439	44.2	2.89
2	333	25.5	2.27	279	39.4	2.31
3	243	24.3	2.34	236	41.1	2.36
4	191	20.9	2.35	187	35.3	2.65
5	156	20.5	2.09	139	37.4	2.31
6 to 7	166	17.5	1.76	204	22.1	2.38
8 to 10	152	15.8	2.38	187	25.1	2.45
Over 10	83	4.8	1.00	177	13.0	2.30
Total	1717	22.3	2.25	1848	34.3	2.48

bred for the first time in the previous year to only 13% in those which had already bred more than ten times. The comparable figure for males changing sites was a decline from 28% to 5%. Age for age, about half again more females than males changed sites between years (Figure 8.26, Table 8.10). This progressive decrease in changing sites with breeding experience is probably influenced by the order of return to the colony each year, with the oldest birds returning first and so having first choice of sites, while the youngest returned last, by which time sites were already occupied and aggressively defended.

For a time, I thought that when a member of a pair died, the widowed individual selected a new mate from similarly affected birds of the other sex within the colony as a whole, but this proved not to be the case. The choice of partners was much less than would be expected from the numbers of the other sex who also required new partners from within the colony as a whole. For example, in a colony of 80 pairs, the partners of about 13 males will have died since the past year, and there would be about 16 females who had lost their mates. It is now evident that the males do not select a new partner from the available 16 but from, perhaps, only two or three widowed birds and a similar number of divorced females, all of whom had nested near the male in the previous year. The choice of a new mate was much restricted.

There is an alternative option for a male with previous breeding experience seeking a new partner, which is to attract a female that has not bred before and did not have strong attachment to a particular site or part of the colony. In fact, 12% of males who had bred before took a female partner who was breeding for the first time (see Appendix 1).

The great majority of birds of both sexes that changed their nesting sites moved only short distances between sites. Overall, the distances moved averaged 2.23 and 2.48 sills (3.1m and 3.5m) for males and females respectively, which is much less than if the new site was selected at random within the colony (which, on average, would have involved moving about six sills, equivalent to moving 8.4m) (Figure 8.27).

There were several situations which caused a different nest site to be used. For example, if both members of the pair returned, but took different partners, then one or both had to change nesting site. Again, if one of the partners had died since the previous breeding season and the surviving bird obtained a new mate with previous breeding experience, one of the new pair had to be using a different nest site.

In general, when both members of the pair divorced or the female partner died, the male retained the nest site. In the case of divorce, the female moved elsewhere. When the male of the pair from the previous year died, the response of the female was variable, but most frequently the surviving female moved site and another pair took over the site which she had used in the previous year. In contrast, it was typical of old females to retain the site when their partners died. This created a problem, since most older males also retained their site and attracted a female to nest there. This resulted in a quarter of females which were breeding for at least the tenth time forming a new pair with young males who were about to breed for the first, second or third time (Appendix 1).

The degree of selection involved in acquiring a new partner is much less than might have been expected, but the short distances moved maintained the *status quo*, keeping edge birds to the edge and central birds continuing to nest within the centre of the colony.

Overall, 71% of females that changed site also took a new mate, and this proportion changed only marginally and not meaningfully as age increased

FIG. 8.27   *The distances between nest sites used by male and female kittiwakes in consecutive years. The average proportion of other sites available from each nest site in relation to distance is shown by the dotted line and the difference of this from the observed pattern of change of site clearly demonstrates the reluctance of kittiwakes to move far from their previous nest site.*

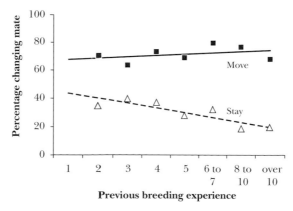

FIG. 8.28 *A comparison of the percentage of female kittiwakes of different breeding experience that moved site and changed mate (squares) compared with the percentage of those which retained the nest site and also changed mate (triangles). The proportion of those that stayed changed meaningfully with the age of the females, but for those which moved it was unrelated to age.*

(Figure 8.28). However, 29% of females that moved sites also remained with their partner of the previous year. In some cases, pairs had been re-established on the site used in the previous year, but were subsequently ousted from the site by another pair and moved elsewhere together. In other cases, both

FIG. 8.29. *The Brewery Warehouse at North Shields in 1976. The kittiwakes nested on all four sides of the building, including the side out of sight of the river. The original nests were built on the ledge of the top floor door, on the river side of the building.*

moved spontaneously to another site without first reforming the pair at the original site. It is evident that members of pairs recognised each other as individuals by voice (Wooller 1978, Mulard *et al.* 2009) and probably other characteristics which the human eye cannot easily detect, thus facilitating moving the nesting site some distance while still maintaining the same partnership.

Among those females that returned to the same nest site, the proportion which had also changed mate was much lower, and this decreased from 40% to 20% as their breeding experience increased (Figure 8.28). The low level of about 20% in the oldest females could be attributed almost entirely to the proportion whose partner had died since the last breeding season and confirms the extremely low level of divorce among these older birds.

It is now evident that there are a series of factors which bring about changes in the timing and success of breeding of individuals within a kittiwake colony. The most important of these are age (expressed by years of breeding experience), social behaviour effected through neighbouring pairs and the tendency for the more productive individuals to occur at or near the centre of the colony.

CHAPTER 9

Long-term changes in individual colonies

WHEN making casual or occasional visits to a kittiwake colony, it is rare to see much change in the biology of the birds between successive years. It is only with frequent visits and the recording of quantitative information over many years that trends become evident, and considerable changes do occur over time. For example, the date of return to the colony changed, as did the start of laying, clutch size, breeding success and adult survival rates – and all have been recorded and documented from the systematic collection of information over many years. The longest run of information on kittiwakes relates to the North Shields colony in north-east England, where data were collected annually from 1953 to 1990, and at nearby Marsden where data collection started in 1953 and continued intermittently until 2010.

Information for shorter periods of time is also available for breeding success at several other colonies in the UK, such as Foula in Shetland, the Isle of May in Scotland and the Farne Islands in England, and in other countries, including northern Norway and at Middleton Island in Alaska. This account is primarily based on the North Shields colony.

CHANGE IN TIME OF REOCCUPATION
OF THE COLONY

Many kittiwake colonies are not easily observed in winter and early spring, and there is relatively little and only casual information on the time of return and reoccupation of nesting sites in most parts of the kittiwake's geographical distribution. Most published accounts indicate that the time of the annual return to colonies is later in more northern areas, but detailed information is rare. Kittiwakes in most of Greenland do not visit nesting cliffs until April, although this can occur in late March in areas of open water in south-west Greenland (Salomonsen 1967). In Arctic Russia, Belopolskii (1957) hinted that the return is in late March or early April in the Murmansk area of Russia and even later further north in Novaya Zemlya.

In Britain, *The Handbook of British Birds* (Witherby *et al.* 1943) states that kittiwakes start visiting nesting cliffs at the end of February in the south of Britain and in the second week of March (occasionally from the third week of February) farther north, and that 'birds settle in from the end of March to mid-April'. *Birds of the Western Palearctic* (Cramp & Simmons 1983) indicates in the phenology diagram that return to the colony starts at the beginning of March, but no further information is given.

In the early months of the 1950s and 1960s, I had the opportunity to regularly visit several colonies in north-east England and some, such as Inner Farne, did not have kittiwakes on the nesting sites until the first few days of March. In contrast, kittiwakes first returned to the nesting ledges at North Shields on 15 December 1977, and after that date the colony was occupied daily until the autumn of 1978. At nearby Marsden, some adults were back on nesting sites during January of every year between 1952 and 1970. In contrast, from 2004 to 2009, the first returns there did not occur until March.

Kittiwake colonies within the same area often show characteristic differences in the time of first reoccupation, and once a colony was initially reoccupied, birds visited every day apart from highly exceptional occasions, such as during severe gales or heavy snowfall. Interestingly, this pattern is very different from that of the Fulmar, where cliffs are periodically and frequently vacated for several days following the first return in November (Coulson & Horobin 1972), and the birds appeared to be using strong winds as a means to reach distant feeding grounds.

What I did not expect to find were large and progressive differences in the time of return at individual colonies as years went by, and yet this occurred at North Shields between 1952 and 1990 (Figure 9.1). These changes formed a long-term pattern, first with a progressively earlier return for over twenty years and then, from 1978, an appreciably and progressively later return and reoccupation of the colony.

The later returns after 1978 also occurred at the nearby colonies at Marsden, and this later return (without exact dates) has also been commented upon at other colonies on the east coast of Britain and elsewhere, and so it

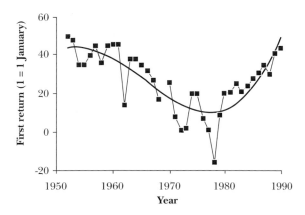

FIG. 9.1   *The date of the annual reoccupation of the North Shields kittiwake colony from 1952 to 1990. The trend line is a third-order polynomial fitted to the data points. Note the December return in 1978. In all years, the first occupation was followed by a progressive build-up in the numbers of birds present and temporary desertions rarely occurred, apart from the colony being deserted each night.*

was a general trend affecting kittiwakes over appreciable areas. It is not clear why these changes occurred, but it leads to suspicion that all is not well in the oceanic wintering areas. Perhaps the kittiwakes are encountering difficulties in building up their body condition during the winter, prior to having to cope with the much more restricted feeding range imposed by becoming attached to a colony, to which they make daily visits.

The pattern of earlier return from 1952 to 1975 at North Shields (Figure 9.1) did not occur nearby at the longer-established Marsden colonies, nor at other colonies in north-east England, and this pattern was peculiar to North Shields. Since the colony was first established only in 1949, presumably by young birds, the progressively earlier return to the colony probably involved two factors – the progressive increase in size of the colony and the increase of older birds, which return earlier than younger individuals. In addition, there is a behavioural prerequisite to reoccupation, in that a group of birds returns to the area and forms a raft on the sea, from which birds visit the colony. It is likely that the critical size of the raft is reached later in a small, new colony than in a larger, well-established one.

LONG-TERM CHANGES IN THE DATE OF LAYING

1. First egg to be laid in the colony in each year (Figure 9.2)

In most years, the first egg in the North Shields colony was laid during the first week of May, but laying occasionally started in the last days of April. The first egg date advanced between 1953 and 1965, and then became later in

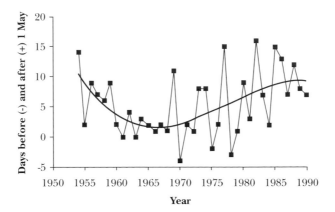

FIG. 9.2 *The date of laying of the first egg each year from 1953–1990 in the kittiwake colony at North Shields. A third order polynomial has been fitted to the data to indicate the general trends.*

more recent years, but the magnitude of this change was only a few days. In all but one year, another female started to lay within two days of the first egg being produced in the colony, so these first dates were not exceptional outliers. Because similar numbers of pairs were studied, the influence of sample size has had little effect on the data, and certainly not after 1959. Between 1966 and 1990, the start of egg laying has tended to be later by about a week. Wanless *et al.* (2009) found a similar trend of a later start of laying in recent years from data collected by the wardens on the Farne Islands, about 40km further north.

2. *The mean egg date in the colony (Figure 9.3)*

The average date on which the first egg of each clutch was laid in the colony showed a similar pattern to the date of the first egg laid in the colony each year, with an initially earlier average date until about 1965, and then slightly later dates in more recent years. The changes are less pronounced than those of the first egg date and, for example, the date changed by less than three days between 1966 and 1990 compared with a week for the first egg.

3. *Last egg to be laid in the colony (Figure 9.4)*

The date of laying the last clutch to be started in a colony each year is a parameter which is rarely collected, yet it can have considerable significance. At North Shields, the date of the start of the last clutch each year did not show consistent changes over the 37 years of study, and averaged about 9 June. There were irregular fluctuations of up to 15 days between extreme dates within the whole study period, and one source of this variation was the small size of the colony in the early years, which made an extreme date less likely. A second cause was the less accurate estimated date of the last egg because it

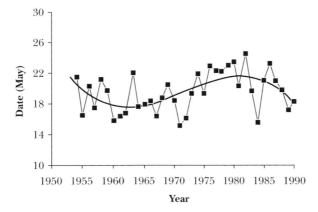

FIG. 9.3 *The mean date of the start of all clutches laid by kittiwakes at North Shields in each year, 1954–1990.*

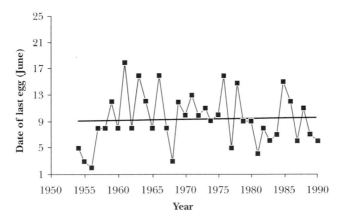

FIG. 9.4 *The date on which the last egg was laid in the kittiwake colony at North Shields, 1954–1990. The best-fit line to the data indicates that the end of egg laying did not change over the period of study.*

was usually a single-egg clutch; based on our usual two visits per week, there was a potential error of up to two days. Nevertheless, it is evident that the end of laying each year showed no long-term trend and remain unchanged throughout the study. However, it should be noticed that in most years there were pairs which built nests and mated, but failed to lay eggs, so the last clutch date was not the result of there being no more birds which could have laid. It is relevant to note that the few pairs which lost their clutches earlier in the breeding season did not relay after this date. These observations suggest that an environmental factor which did not change over the study period (perhaps linked to photoperiod) inhibited egg laying after this date.

4. Spread of laying dates within the colony (Figure 9.5)

From 1954 to 1961, the spread (in days) between the start of the first and last clutches in the North Shields colony increased significantly. This was probably associated with the progressive increase in size of the colony over this period, making extremes more likely. When this value was corrected for the size of the colony, the trend virtually disappeared. From 1960, the number of pairs breeding remained at a similar level, and so from that year, changes in the size of the colony cannot explain the progressive decline in the spread of breeding in later years.

This reduced spread of laying was caused simply by the later onset of laying in the colony, but it resulted in breeding within the colony becoming progressively more synchronised, although there is no evidence that any advantage was obtained by this. In fact, greater synchrony was associated with a slight decrease in the annual clutch size. Increased synchrony in breeding was also evident when a better measure of the degree of synchrony, the standard deviation of the laying date each year, was used.

LONG-TERM CHANGES IN CLUTCH SIZE

The average annual clutch size at North Shields tended to decline over the study period, but there was considerable year-to-year variation. The trend is better explained by a meaningful second-order polynomial curve (Figure 9.6). From 1954 to 1968, the average annual clutch size in the colony each year was two eggs or greater. This was exceptional, as in most kittiwake colonies in Britain the annual average clutch has been consistently slightly less than two.

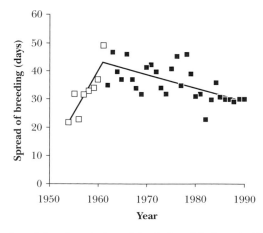

FIG. 9.5   *The number of days from laying of the first and last egg in the kittiwake colony at North Shields, 1954–1990. Two periods are recognised, namely that caused by the growth of the colony from 1954 to 1960 and the subsequent period when the size of the colony was virtually constant and the spread of breeding progressively decreased.*

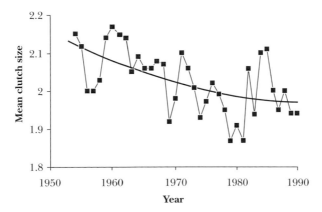

FIG. 9.6   *The mean clutch size laid each year by kittiwakes at North Shields, 1954–90. The second-order polynomial curve is a better fit to the data points than a straight line but both indicate a meaningful trend towards a lower clutch size during the study period.*

The average clutch at North Shields was higher in years when the average date of laying was earlier (r_{35} = −0.42, P < 0.01) and so the unusually high average clutch sizes in earlier years are probably the result of the earlier onset of laying at North Shields than in other colonies. Similarly, the lower clutch size in the later years at North Shields may have been caused by the later onset of breeding in more recent years.

LONG-TERM CHANGES IN HATCHING AND BREEDING SUCCESS

The percentage of eggs that hatched and the percentage of hatched chicks that fledged were recorded each year (Figure 9.7). Levels of success changed little over time, with the most characteristic effect being that fledging success was appreciably higher than hatching success in every year except 1954 (which was based on a small number of nests). There is an indication that both values tended to increase marginally during the 1950s and again in the 1980s, with a hint of a slight dip in the 1970–80 period, but the values were generally high throughout the study period. There were no years with very low hatching and fledging successes. It is particularly important to recognise that the high fledging success indicates that food shortage was not an important factor in the breeding performance of the North Shields kittiwakes during the entire study period.

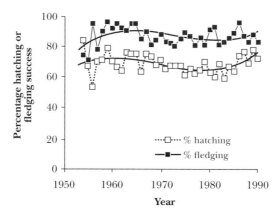

FIG. 9.7   *The hatching success and fledging success of kittiwakes nesting at North Shields, 1954–1990. Note that the fledging success was higher than the hatching success in every year except one. Third-order polynomials have been fitted to the data points.*

Long-term changes in the number of young fledged per pair (productivity)

These values (Figure 9.8) are the product of the hatching and fledging successes at North Shields, which showed a small increase in the 1950s and 1980s, with a slight dip in the 1970s. However, the variations are small, and in every year an average of 0.95 or more young were fledged per pair, indicating successful breeding in every year. Successful breeding also continued at nearby Marsden for the next 20 years (1991 to 2010), with the exception of 1998, when many adults were killed by a neurotoxin during the breeding period.

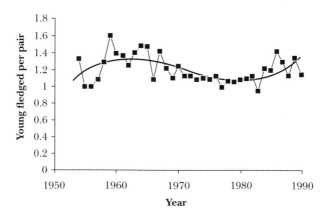

FIG. 9.8   *The number of young fledged per breeding pair (productivity) of kittiwakes at North Shields, 1954–1990. A third-order polynomial has been fitted to the data to indicate the underlying trends.*

LONG-TERM CHANGES IN ADULT WEIGHT

Each year, adult kittiwakes at North Shields were captured, colour-ringed, weighed and then sexed by their behaviour or biometrics. Most were weighed between May and early July, during active breeding. These measurements produced a long series of over 1,500 weights recorded from 1953 to 1990 (Figure 9.9). There were no long-term changes in weight of either sex, with the males averaging 387g and females 343g. There was a tendency for the relatively small differences in particular years to occur in the same direction for the sexes (confirmed in Figure 9.10), and this suggested that weight was

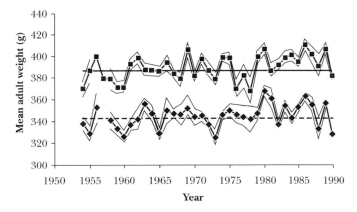

FIG. 9.9   *The weight of male (squares and upper lines) and female (diamonds and lower lines) kittiwakes at North Shields, 1954–1990. The thick horizontal trend lines indicate the lack of long-term trends over the study period. The thin lines indicate one standard error about each point.*

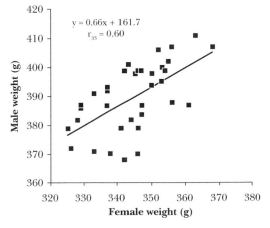

FIG. 9.10   *The correlation between the average weight of adult male and adult female kitti-wakes breeding at North Shields for each year, 1954–1990.*

indeed higher in some years than in others and the correlation indicates that about 36% of the variation in weight occurred simultaneously in both sexes. However, it is reasonable to conclude that the differences of only a few grams, which never exceeded 6% of average body weight in any year, did not have any appreciable effect on breeding performance.

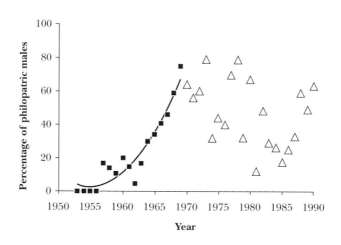

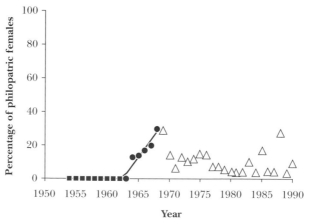

FIG. 9.11   *The long-term trends in the extent of philopatry of male (upper) and female (lower) kittiwakes at North Shields. Note the difference in philopatry between the sexes and the lag time before the first individual reared in the colony returned and bred there. The period from 1954 to 1968 has been represented by different symbols and indicates differences in the pattern of philopatry in the two periods. The earlier period included the time that the colony was increasing and progressively more young were fledged in the colony. Thereafter the degree of philopatry varied markedly from year to year, but showed no meaningful trend, being consistently and appreciably lower in females than in males.*

LONG-TERM TRENDS IN THE EXTENT OF PHILOPATRY

In the past, it was believed that young produced in a colony often, or even invariably, return and breed there. The extent of philopatry in kittiwakes – the return of young to breed at the natal colony – has become increasingly questionable. Through intensive ringing, this has been shown to be less frequent than previously thought. For the kittiwake, this topic has been examined in detail (Coulson & Coulson 2008) and it was established that few females and only a distinct minority of males returned to the colony in which they had been reared. The implication is that the great majority of kittiwakes which breed in a colony have been reared elsewhere, and this was the case at North Shields. Figure 9.11 shows the proportion of the new breeding birds at North Shields which had been hatched there. The two graphs have been put on the same scale to emphasise the difference in the behaviour of the sexes. The high proportions of males among the philopatric individuals during five-year periods are presented in Figure 9.12, and show that this remained high and with no evident trend throughout the whole study period.

In a new kittiwake colony, it is several years before the relatively few chicks reared there are old enough to breed, resulting in a time lag of several years before the first cases of philopatry could occur. The level of philopatry was initially low, but increased progressively (Figure 9.11). That increase did not continue and, in the next twenty years from 1970, the level of philopatry fluctuated irregularly around 50% in males and below 10% for females. There was no indication that this proportion would change in older colonies.

THE SURVIVAL OF IMMATURE KITTIWAKES

An index of the survival of young from each year class produced at North Shields was made from the extensive sightings of marked birds subsequently seen at North Shields and in other colonies. These estimates could not

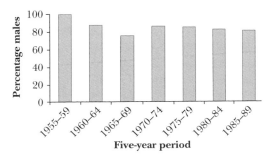

FIG. 9.12   *The percentage of males among all philopatric recruits to the North Shields colony in five-year periods. Based on 288 philopatric kittiwakes. Only eight birds were available in the 1955–59 period and the remaining time periods each had at least 31 philopatric birds.*

include all of the young which had survived, because many of the young marked at North Shields moved elsewhere to breed, and so outside the area regularly searched. This meant that only an index which could be used in a comparative way was obtained, and the real survival rates must be higher. The records used came from sightings at colonies by many people, including those actively involved in the research at North Shields, and also from ringing recoveries of the marked birds. The intensity of observations could not be confirmed as having remained the same throughout the study period, but observer effort did not change markedly and rapidly during the study, and the high proportion of birds subsequently recorded was reassuring in that most differences found in the survival between adjacent cohorts were real. The criterion used was that evidence existed that the individual had survived for at least two years.

Young birds reared at North Shields were originally marked with a single colour to denote the year class and a numerical BTO standard ring. Each of these individuals had to be identified by reading the BTO ring number through a telescope, and this was often difficult to achieve. From 1972, a colour ring engraved with a unique combination of two alphanumeric symbols was used and this allowed individuals to be identified from a greater distance and more quickly, resulting in a much higher identification rate. Birds contributed to the index if evidence was obtained from any source that they had survived for two or more years after hatching. The overall resighting rates of those marked from 1972 to 1990, using the alphanumeric rings, were

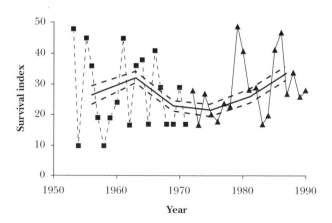

FIG. 9.13   *The percentage of fledged kittiwake chicks known to be alive two years later, used as an immature survival index. The data from 1972–1990 are based on sightings of young which had individually numbered colour rings. The data from 1954–1971 are based on birds which were identified by reading metal ring numbers and scaled up by 2.4 times to the same average as 1972–1990 to achieve comparability. A trend line ± SE is based on six-year averages. Note there was appreciable survival from every year class, but there were some years when survival was appreciably higher, in particular 1979, 1980, 1985 and 1986 (more recent years), while survival in the whole period 1967 to 1978 showed little variation and no exceptional years.*

2.4 times greater than those from 1954 to 1971. Accordingly, the earlier data were increased by 2.4 times to give comparable overall percentage values for the early and later years. Figure 9.13 shows marked year-to-year fluctuations, and these represented differential survival of particular cohorts of young birds. Some young survived from all cohorts and there were no years where there had been total failure to produce young which survived to adulthood. The trend line based on six-year averages indicates that survival of the immature kittiwakes was, in general, lower from cohorts produced from 1967 to 1978 than at other times, and this was mainly the result of the absence of exceptionally good years during this period. It is interesting to note that while the annual adult survival rates decreased progressively over the whole period, there was no similar trend in the index of survival of immature kittiwakes, and the index actually increased from 1972 to 1990 while adult survival declined. It appears that the factors influencing the survival of adults were mainly different from those which affected immature kittiwakes. There were four years where the uncorrected data (i.e. after 1971) recorded sightings of over 40% of the fledged chicks, known to be still alive two years later. The actual survival over the two years in these cases must have been very high.

MISSED BREEDING YEARS BY ESTABLISHED BREEDERS

The percentage of birds which had bred in a previous year and would breed again in a future year but missed a breeding season is shown in Figure 9.14. Because of the low proportion of adult birds missing a breeding season, the data have been grouped into four time periods. Overall, females miss a breeding season about half again as frequently as males and this may reflect

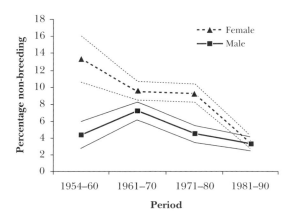

FIG. 9.14 *The percentage of female (triangles and dashed line) and male kittiwakes (squares and thick line) missing a breeding season in four periods of the study. The thin lines show one standard error from the means.*

the higher mortality rate in males and the greater likelihood that there would be a surplus of females. There was a trend for a smaller proportion of adults to miss breeding in the later years and this was particularly evident for females. The reason for this trend is not obvious, but it may indicate that there was less severe competition for mates or that the adults were in better condition in later years.

COLONY SIZE

The colony at North Shields was founded in 1949 and the number of nests increased rapidly and progressively to 104 nests (pairs) in 1964. This was followed by a brief plateau, and then a small decline to 1970. From 1970 to 1990 the numbers of nests each year fluctuated between 92 and 64 (see Figure 9.15). The relative stability in numbers in the last 20 years of this colony is of considerable interest, because it occurred during the same period in which the annual adult survival rate decreased by an average of ten percentage points (see below). It is evident that the recruitment to the colony changed and was able to compensate for the greater mortality.

ANNUAL ADULT MORTALITY RATE

The long-term effects of survival and mortality are closely related (percentage mortality rate = 100 – percentage survival rate) and the mortality rate is presented here because it reveals the changes more clearly. Although the mortality rate of breeding kittiwakes at the North Shields colony fluctuated from year to year, the data shown in Figure 9.16 clearly show that there was an

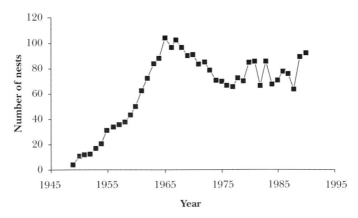

FIG. 9.15   *The number of nests in the North Shields kittiwake colony from 1949 to 1990. Note the almost exponential increase to about 1965 and then the numbers fluctuated around 80 breeding pairs thereafter.*

underlying progressive increase in mortality which extended for the whole of the study. The increase in mortality rate was five percentage points over each ten-year period and involved a threefold increase in the mortality over the study period from 1954 to 1990. This was an appreciable change.

The colony ceased to exist in 1991, but mortality rates are also shown for a marked group of birds within the nearby colony at Marsden. These show that the declining mortality rate in the late 1980s at North Shields continued in the 1990s, until a period of exceptional mortality occurred in July 1997 and continued during 1998, which decimated the colony and killed most of the marked birds that had been used to calculate adult mortality and survival rates.

RECRUITS TO THE BREEDING GROUP

The numbers of new breeding birds which began to breed at North Shields each year varied considerably over the study period (Figure 9.17). During 1954–65, the numbers of recruits increased as the colony also increased rapidly in size (giving the large standard error). From 1966 to 1975, the annual number of recruits was low but progressively increased in the later years, from 1975 to 1990, and compensated for the increasing annual adult mortality rates. Why the numbers of recruits each year decreased after 1965 is not clear, but in part it may be associated with the majority of suitable nesting ledges having been occupied by that time. However, it coincided with a period of lower survival among immature kittiwakes (see Figure 9.13).

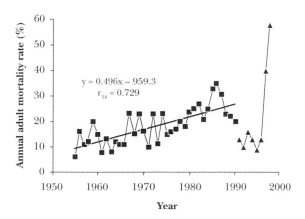

FIG. 9.16   *The annual adult mortality rate of kittiwakes breeding at North Shields (squares) 1954–90, and the subsequent mortality rate nearby at Marsden (triangles), 1991–98. The trend line shows the progressive increase in the mortality rate at North Shields, which increased by about one percentage point every two years. The mortality rate declined in the 1990s, until a mass mortality of adult kittiwakes in the area occurred in 1997 and 1998.*

Age at first breeding

The age at first breeding was not known for most of the kittiwakes at North Shields, because many of the males and almost all females had been reared in other colonies where the young were not ringed. While some of these immigrants showed signs of having retained a few feathers from the immature plumage, these proved not to be totally reliable as an indicator of their age, since a few three-year-olds ringed as chicks showed no immature feathers. Of the 263 birds which were ringed as chicks, and so were of known age when they bred for the first time at North Shields, all but ten had been originally ringed in the North Shields colony. Table 9.1 shows the average age of these birds in three ten-year periods during the study. Useful data did not exist from 1954 to 1960, as only two birds ringed as chicks were recruited in this time period.

There was a highly significant decrease in the age at first breeding for males in the later years, while the much smaller sample of females was less convincing, but they did show a meaningful decline in age between 1961–1980 and 1981–1990. The question remains as to whether these philopatric individuals, which of course bred in their natal colony, bred at the same age as those which immigrated to North Shields from elsewhere.

General comments

Many parameters recorded at the kittiwake colony at North Shields during the period 1954 to 1990 showed little or no change over time. Some showed consistently high levels, such as clutch size and productivity, and consistently low levels of philopatry, particularly in females. Where changes were detected

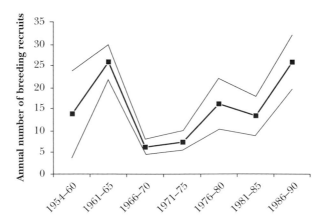

FIG. 9.17   *The annual number of recruits to the breeding group of kittiwakes at North Shields expressed as an annual average. The thin lines represent one standard error from the five-year average periods.*

TABLE 9.1 *The age at first breeding of male and female kittiwakes which had been ringed as nestlings and later nested at North Shields.*

Period	No. of males	Mean age of males (years)	No. of females	Mean age of females (years)
1961–1970	64	4.59 ± 0.12	17	4.22 ± 0.26
1971–1980	69	4.42 ± 0.12	15	4.25 ± 0.19
1981–1990	77	3.69 ± 0.10	21	3.61 ± 0.22

between 1954 and 1990, they were mainly adverse, but most had only a small effect on breeding (Table 9.2). The major changes occurred in the date at which the first birds returned to the colony each year, the onset of laying each year, the age at first breeding and the increasing level of mortality among the breeding kittiwakes.

Productivity

The high fledging success, high productivity and the consistent weights of adults all indicate that during the breeding season, there were neither frequent nor severe food shortages. In every year, many kittiwake pairs were successful in rearing two chicks to fledging, and it is worth noting that this high productivity continued in the nearby colonies until at least 2010. Throughout, the main food of kittiwakes in the study area was sandeels, and it is evident that the density and availability of sandeels within the feeding range of the kittiwakes have been more than adequate since 1954. However, this is not a conclusion that can be applied to all colonies in Britain and elsewhere, where varying frequencies of poor breeding performance have been reported in more recent years, examples of which are shown in Appendix 3.

Sandeels (or sandlance) are the most frequently taken food by breeding kittiwakes in Britain, although there are exceptions (see Chapter 2), and there is evidence that their stocks have varied dramatically in some areas in recent years, resulting in poor breeding success among kittiwakes at several colonies. Oro and Furness (2002) have produced convincing evidence of this for the area of Shetland around Foula, which produced marked fluctuations and virtually total failure in the production of young kittiwakes in recent years. Similar failures to breed successfully have been recorded from other colonies (Appendix 3). These contrast markedly with the situation at North Shields and Marsden where productivity remained invariably high. Whether these fluctuations in sandeel abundance are natural or caused by intensive fishing is not clear. It has been stated that sandeel stocks have been drastically reduced by commercial fisheries, producing breeding failures in kittiwakes. Monitoring sandeel abundance and the quantities of sandeels taken by commercial trawlers has not been adequate in the past and until the last few years it has been difficult to form a clear quantitative picture of the impact of such fisheries. What is without doubt is that harvesting sandeels as a means of producing fishmeal for farmed animals, fertiliser and oil is short-sighted.

TABLE 9.2 *The main changes in the breeding of kittiwakes at the North Shields colony, 1954–90.*

Breeding parameter	First adverse appearance	Last adverse appearance	Peak of adverse effect	Maximum size of effect
Return to colony	1979	1990	1990	About 50 days
Laying date	1974	1989	1976–82	Delay of *c*.6 days
Clutch size	1969	1990	1979–81	Decrease of 9%
Hatching success	1973	1983	1975–83	Decrease of 10%
Percent fledging success	1966	1986	1973	Decrease of 10%
Percent breeding success	1966	1983	1973–78, 1982–83	Decrease of 17%
Young fledged per pair	1966	1983	1977–83	Decrease of 18%
Adult mortality rate	1967	1990	1984–87	Rate doubled

Sandeels are the main food of many marine birds, seals, porpoises and larger commercial fish, such as Cod, in waters off the coasts of western Europe and in many other areas. To threaten the existing stocks for non-human food by intensive commercial fishing is not a sensible act and some, but by no means all, of the sandeel fisheries around the UK have been banned. The breeding performance of kittiwakes is indeed an environmental indicator of the condition and well-being of the marine environment, but what has actually changed in the environment is much less clear!

While sandeel numbers have fluctuated in many areas, the abundance in the area used by North Shields kittiwakes has always been high enough to prevent years of poor breeding. Should availability of sandeels there temporarily decline, there are other fish, including sprat and small gadoids, which could be used as alternative food. The situation in Shetland is different and there is a lack of alternative food species. Even though Herring are abundant in these waters, they are adult fish and too large for kittiwakes to handle. Small, young Herring are absent and do not reach that area (R.W. Furness, *in litt.*).

Mortality

The doubling of the mortality rate of adult kittiwakes of both sexes during the period 1954–1990 is puzzling. There was no predation of any kind on the kittiwakes at the North Shields colony, except for a single instance when four adults were shot by youths in one year (and excluded from the mortality estimates). There were no birds of prey, such as Peregrine Falcons or White-tailed Eagles, taking adults as prey. Food was not in short supply, as indicated by the consistently high level of production of fledged young year after year. Further, 81% of adults which disappeared, and presumably died, did so between September and March, when the birds were away from the colony, and only 19% disappeared during the breeding season. This high proportion of kittiwakes dying in the winter part of the year, when they are oceanic, probably explains the low ring recovery rate in this species.

The seasonal difference in the number of kittiwakes dying suggests that the conditions in the wintering area were directly or indirectly causing mortality. If so, this raises questions about the oceanic marine environment used by the kittiwakes, about which we know very little. What is killing the kittiwakes when oceanic? In what way have conditions deteriorated during the 1954–1999 period? There is evidence that the death of many adult kittiwakes was preceded by a period of several months of decline in their condition, indicated by poorer breeding performance when they were breeding for the last time. This type of progressive loss of condition before death can be caused by slow-developing diseases, and while disease may be acquired at any time, death is more likely to occur during periods of stress. At North Shields, breeding did not appear to be a period of great stress on the adults. Perhaps the protracted main moult that starts during breeding and continues afterwards is an appreciable stress, particularly when the longest primaries are not shed until the adult kittiwakes are oceanic. The reduced power of the wings as a result of the loss of the longest primaries may well increase the risk to adult kittiwakes when exposed to severe weather conditions while oceanic. It may be of significance that the primaries of young of the year are not moulted and that immature kittiwakes entering their second and third years moult earlier than the adults and have fully functional primaries earlier, often by September, and so before they become fully oceanic.

The mortality of immature kittiwakes did not mirror the increased annual mortality experienced by the adults, yet they spend even more time in a pelagic life. So perhaps the increasing mortality of adults was not simply a direct effect of changes in the oceanic environment. An indirect way of testing whether the oceanic conditions contributed to the increasing mortality is this: years of high and low mortality should coincide in kittiwakes from other colonies because they become mixed in their wintering areas and are likely to encounter the same overall oceanic conditions, while those suffering mortality in the breeding areas might not be expected to correlate.

Currently, such data are very limited and from more recent periods than the North Shields data set, but those from Skomer, in Wales, showed a meaningful correlation between the annual survival rates of breeding adults at North Shields and those on Skomer breeding adults. However the correlations between these data and a shorter series based on the Isle of May with less intensive re-sighting is less encouraging. More evidence is needed, but the decline in many kittiwake colonies suggests that this increased overall mortality may be a general effect and so one of particular concern.

Recruitment to the breeding group

Following the period of rapid colony growth, the size of the kittiwake colony at North Shields went through a period of over 25 years of relative stability in the number of breeding pairs, fluctuating around 80 nests. The recognition of this stability is important, because the progressive increase in the adult mortality rate must have produced an appreciable perturbation, and this was counteracted by progressively increasing recruitment of new breeding birds. If the mortality rate of adults in the 1960s produced stability, then the higher mortality in later years (Figure 9.16) should have caused a decline, unless a 'colony-size' effect was operating to prevent it. The decrease was avoided by an increasing number of new breeding birds attracted to the colony in later years. This, in turn, suggests that the group of birds within a colony might function in a way which tends to produce self-regulation of numbers (homeo-stasis), that is, the colony is capable of regulating its size around a given level and so compensating for variations in adult mortality.

This observed stability would appear to be inconsistent with the virtually constant production of young birds within the colony, and there was little indication that these young birds suffered lower mortality rates in the later years. But this objection only applies if one thinks of a colony as a discrete, self-perpetuating unit, supplied by its own young production, and that is not the case. As explained in more detail in Chapter 11, most of the new breeding birds (particularly the females) which bred at North Shields had been reared elsewhere and immigrated into that colony (while many of the young females reared there moved to breed elsewhere). So there is no need to accept this potential inconsistency. In effect, each kittiwake colony attracts new birds from a pool of young produced in many colonies within 1,000km or so. The nature of this attraction is not totally understood, but one factor is the number of sites within the colony which each year are unoccupied by estab-lished breeders because of mortality. This stability could be produced in three ways. Firstly, when there are more vacancies within the colony, more potential recruits are attracted and remain in the colony. There are more of these vacancies following a year of high mortality, and when there are more unoc-cupied spaces, the greater the attraction for new birds. This increased attractiveness produced by unoccupied sites explains why a new colony tends to grow rapidly. Secondly, the proportion of adults which have bred before and miss a breeding year can act as a reservoir of breeders, and as shown in Figure 9.14 the level of non-breeding has declined as the adult mortality rate increased. Thirdly, potential recruits do not spend as many years visiting the colony prior to breeding because the difficulty of obtaining a site decreases as the adult mortality rates increase. This last possibility would result in the age of first breeding being reduced, and this was indeed the case, with males starting to breed almost a year younger, and females 1.6 years younger, by the end of this study. All of these possibilities indicate a cushion of potential recruits all of which can respond to negate an increased mortality of adults. Of course, if mortality of adult kittiwakes is high over a large geographical area, then there may not be enough recruits to compensate everywhere for this. In the situation of increasing adult mortality rate, the recruits will no long be able to reduce the age at which they breed and the colony size will decline. In other words, variable attraction of the colony to recruits, non-breeding by established breeders, and variation in the age at first breeding, can all act in a density-dependent way and compensate for changes in adult mortality.

Thus the natural perturbation caused by a change in adult mortality has allowed an insight into how a colony functions. V.C. Wynne-Edwards (1962) was correct to suggest that that the control of recruitment of breeding adults was a function of colonial breeding. However, he failed to consider the important fact that regulation of colony size does not necessarily regulate the size of the population over a large geographical area, as more (or fewer) colonies could be created (or lost). The dramatic increase in kittiwake colonies within the North Sea area during the past hundred years is a testimony to this. Colony size regulation is not population regulation.

CHARACTERISTICS OF NEW COLONIES

It seems likely that some of the changes observed at the North Shields colony up to about 1960 were effects associated with it being a new colony, and similar effects have been observed in a series of new colonies which I have been able to study or were recorded by others. These characteristics of new colonies are listed in Table 9.3.

New colonies are typically small and are usually composed of young birds which return late to breeding sites. Further, the development of the critical numbers of birds in a raft off the colony before it is first occupied each year takes longer to accumulate and involves a higher proportion of the birds associated with that colony. As a result of these factors the first occupation each year of the new colony is late. The small size also results in the extremes of laying being reduced and so the dates of the first and last eggs to be laid in a breeding season tend to be less extreme, resulting in an apparent shorter spread of breeding. Because of this size effect, the standard deviation of the date of laying is a better measure of the spread of breeding.

New colonies typically grow rapidly (Coulson 1983) and often double in size annually for the first two or three years, but thereafter increase at a progressively lower rate. Their initial growth for the first ten years or so is almost entirely dependent on successfully attracting immigrants because potentially philopatric individuals have not reached breeding age for four or five years and in any case the number of young produced in the first few years of the colony is few. Presumably new colonies are attractive to potential recruits because there are numbers of unoccupied nesting sites available within or at the immediate edge of the colony.

On some occasions, new colonies are formed by established breeding birds transferring from a colony which has become unfavourable (Danchin 1992), but much more usually they are created by young birds which have been present at the new site one or more years before the first eggs are laid. However, it should be noted that while the rate of growth of a new colony is high, the number of new breeding birds forming new colonies in relation to

TABLE 9.3 *Characteristics of a new kittiwake colony based on the early years of breeding in the North Shields colony and confirmed by observations on other new colonies.*

Characteristic	Effect 1953 to 1962	Comment
Time of return	Became 25 days earlier	Age and threshold effect
Time of first laying	Became about 15 days earlier	Density and age effects
Average laying date	About fives day later	Became progressively earlier
End of laying	Became about 10 days later	Effect of small number of pairs
Spread of breeding	Increased from 20 to 45 days	Effect of small number of pairs
Clutch size	Little effect	Close to that expected for date of laying
Growth of colony	Rapid. Increased from 20 to 80 pairs	In terms of rate, the most rapid time of growth
Breeding success	Increased marginally	
Productivity	Increased marginally	
Age composition	Increased	Most recruits were young birds

the number available in the population as a whole is relatively low, and over 95% of all recruits select older, established colonies.

In the study area, adults which had bred before invariably returned to the same colony in successive years and most, if not all, birds in a new colony were young. The age composition grows progressively as birds age and return there in subsequent years.

CHAPTER 10

Kittiwake survival rates, longevity and causes of mortality

SEABIRDS are typically long lived, and the kittiwake is no exception, with individuals known to have lived for over 28½ years in the wild. It seems likely that even older birds will be found, particularly in the Pacific Ocean where adult survival rates seem to be higher (see below).

Longevity is not a very useful parameter because it conveys little of what happens to individuals between birth and death; it simply measures the extreme life span of some individuals. Because of this, other measures are used, the most common being the average annual survival rate. This can readily be converted to the annual mortality rate (percentage mortality rate = 100 – percent survival rate). Another useful measure is the adult expectation of life, which indicates the numbers of years that, on average, an individual will live and breed.

Survival rates in humans are calculated from records of births and deaths compiled by governments, and a vast amount of information is available from many countries for over a hundred years, and from these the proportion of individuals with different life expectancies can be obtained. These studies are used extensively in insurance and in planning government policies, and the methods of calculating survival parameters were eventually applied to other species.

Trees produce annual rings which can be used to determine age. In the case of some mammals, similar rings occur in the dentine of their teeth; shells of some molluscs show annual bands and fish can be aged using annual growth of scales. From such characters, age distribution can be determined. But there is no similar way of determining the ages of adult birds, including the kittiwake, unless the young are marked in such a way that they can be individually recognised in later life and their age determined.

Bird ringing (called bird banding in North America) is the main method of such marking, where a numbered ring is put on a bird's leg. Survival rates have been calculated for many bird species from individuals ringed as nestlings, using the sample of those found dead by the general public and reported to the ringing scheme coordinators. Initially, Margaret Nice (1937) and David Lack (1943) pioneered these methods of measuring survival in the mid-20th century. The former collected information on the lifespan of individual Song Sparrows by observing and identifying ringed individuals, while the latter obtained similar information on Robins (European) from recoveries of British ringed birds reported when they died. Information collected at the time of these studies indicated that, in both species, about half died each year. This was received with much scepticism, because it was commonly known that small birds lived much longer than that in captivity, but their estimates proved to be correct.

In the case of the kittiwake, Lack's method of analysis is not very useful for several reasons. The recovery rate (the proportion of birds ringed and subsequently reported, usually dead) is low and effectively only 1.6%. The recovery rate is low for several reasons, but mainly because kittiwakes are pelagic in winter and feed offshore in the breeding season, with the result that many individuals die far from places where people are likely to find them. Individuals that die at sea decay and sink in a few days, long before they can be washed ashore.

Another problem was that rings used until 1960 were made from aluminium or aluminium alloys, and those used on kittiwakes soon became illegible and even fell off within four years, which in many cases was long before the birds died (Coulson & White 1955). Aluminium coming into contact with seawater leads to corrosion. It was not until the 1960s that durable rings made of alloys of copper and nickel (called 'monel') and of nickel, iron and chromium ('incoloy'), were introduced in Britain. Elsewhere, stainless steel rings were introduced at a later date. The durability of these rings was much improved and they showed very little signs of wear after 20 years on kittiwakes; clearly they would last for the lifetime of kittiwakes. To calculate a reliable survival rate without making additional assumptions, we can use only recoveries of young birds which were ringed long enough ago so that all chances of any birds being still alive and perhaps producing further recoveries have passed, that is, restricting the recoveries used to birds initially marked some 30 or more years earlier. Currently, only those ringed between 1960 and 1980 are useable and these are too few to obtain realistic estimates of survival rates. Yet another limitation of using ringing recoveries is that the sex of kittiwakes is unknown when they are ringed as nestlings and few are sexed when recovered, so the sex-related survival rates cannot be calculated.

Because of these difficulties, other methods have been used to estimate the survival rate of kittiwakes. These have further developed Margaret Nice's method, and use combinations of colour rings so that individual birds can be identified. The method necessitates locating and recording all of those which survived each year for a number of consecutive years.

Fortunately, all but two kittiwakes of over 1,500 individuals which bred at North Shields either returned there each year to breed or had died, and the accumulation of records was relatively easy, although it took considerable time making repeated observations at the colony each year. In some other colonies, such as those in France, adults often changed colony and searching for all marked birds became a greater task. More recently, statistical methods have been developed which estimate the numbers of birds which were alive, but failed to be recorded in some years. These mark–recapture techniques are used to compensate for situations where less intensive searches are made for the marked birds each year, but of course, there is a greater likelihood of error in these estimates.

Such corrections were not needed to calculate the survival rates of kittiwakes at North Shields, where the intensive recording of individuals (see Appendix 1) ensured that all pairs of breeding birds were identified each year. The presence of some individual breeding birds was recorded up to 30 times in a single year and their identification was greatly assisted by every breeding bird having a unique combination of colour rings. Rarely, a problem of identification arose (e.g. the rings changing position on the leg, or mistakenly using the same combination of colour rings on two individuals), but the bird's identity could be quickly confirmed by reading the number on the BTO metal ring which could be seen at close distance through the windows of the nesting ledges.

At North Shields, the annual survival rates of colour-ringed adult kittiwakes were determined from 1954 to 1990, and from 1991 to 1999 similar estimates were obtained using marked adults nearby at Marsden. The survival rates calculated were simply the proportion of birds which bred last year that returned to the colony again in the following year.

The information on survival was broken down by sex to give a measurement of the sex difference in the survival rates and it could also be related to the age of the individuals to look for age-related survival rates.

SURVIVAL RATES OVER THE PERIOD OF THE STUDY

The annual survival rate of kittiwakes at North Shields fluctuated over the length of the study, and showed a progressive decrease from 1954 to 1986. During this period the annual mortality rate doubled, but then recovered in the late 1980s and early 1990s (Figure 9.16). In 1998 and 1999, a catastrophic mortality occurred which resulted in the death of most of the colour-ringed birds and ended this aspect of study.

It is interesting to note that despite the decreased adult survival rates over much of the period of the North Shields part of the study, the colony there maintained its size until 1990, suggesting that recruitment compensated for the greater loss of breeding adults. Overall, the annual survival rates of kittiwakes at North Shields and at Marsden averaged 82% ± 1% between 1955 and 1997. Thus, the average annual mortality rate was 18% ± 1%.

OTHER ESTIMATES OF KITTIWAKE SURVIVAL RATES

There are eight other estimates of the survival rates in adult kittiwakes from other colonies, including three in Britain. Studies on colour-ringed kittiwakes on Skomer Island, off the west coast of Wales (C. M. Perrins, pers. comm.) have measured survival for 27 years and these overlap for 19 years with the studies in the Tyne area of north-east England. Of particular interest is that the annual survival rates for individual years correlated between the two colonies (Figure 10.1), suggesting that an appreciable amount of the mortality might occur when birds from the two areas are mixed in their pelagic winter distribution in the Atlantic. The estimates of the annual survival rates of kittiwakes on the Isle of May, Scotland averaged about 89% per year between 1986 and 2000 (Frederiksen *et al.* 2004b), appreciably higher than at North Shields and Marsden over the same time period, and their annual estimates showed no meaningful correlation with the values for Skomer or North Shields and Marsden.

There are a number of estimates of kittiwake survival rates from other colonies, listed in Table 10.1, but these were measured over appreciably shorter numbers of years. The survival rates from the longer studies have been sub-divided for periods of about ten years so that any changes over time in the survival rates at individual sites are evident. The survival rates at all colonies showed considerable variation between years, hence the relatively large standard deviations shown in Table 10.1 (95% of annual survival rates would be expected to fall within the range of ±2 standard deviations of the overall average) and these indicate that the percentage annual survival rates could vary by over 20 percentage points between pairs of years.

None of the European annual survival rates of kittiwakes are as high as those reported from Alaska, but those were obtained over a relatively small

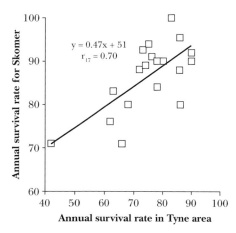

FIG. 10.1 *A comparison of the annual survival rates of kittiwakes on Skomer Island, Wales and at North Shields and Marsden in the Tyne area of north-east England in the same years (1980–1998). About 50% of the variation in survival is common to the two areas. Even without the outlying point of low survival, the relationship remains significant.*

number of years and whether these remain typical awaits information for additional years. Scott Hatch (*in litt.*) has indicated that the survival rates on Middleton Island had probably been reduced in more recent years, but still remained high compared with European estimates. The high survival rates in the Pacific colonies are of considerable interest since the same colonies had exceptionally low breeding success, apparently caused by food shortage. Surprisingly, food shortage during the breeding season which greatly affected the productivity did not appear to have induced lower adult survival in Alaska. Do the adults there abandon breeding in years of food shortage to avoid compromising their own survival?

In Europe, the average annual survival rate ranges mainly between 80% and 88%, although years with survival above 90% were not infrequent and occasional years with survival below 70% have been recorded.

When the survival rate decreased to about 82% at North Shields this did not result in a decline in the numbers of nesting pairs, and the greater loss of adults was compensated by additional recruitment, whereas a similar adult survival rate on Fair Isle was associated with a marked decline in numbers of breeding pairs (Rothery *et al.* 2002). It is obvious that the survival rate of adults is not the sole and consistent determinant of the changes in numbers of kittiwake in colonies, suggesting that the attraction and recruitment of new breeding birds, which may come from colonies over a large area, may be of equal or even greater importance.

TABLE 10.1 *Measures of the annual adult survival rates of the kittiwake. The value of the standard deviation indicates the variation between years and not the accuracy of the mean value. Where records extend over many years, they have been presented in time periods of about 10 years so as to indicate changes in the survival rates found over time. The SD indicates the year to year variation and not the accuracy of the average survival rate.*

Colony	Location	Period	Annual survival rate (%)	SD	Expectation (years)	Method*	Source
North Shields	NE England	1954–60	86.8	4.8	7.1	AC	Coulson & Strowger 1999
North Shields	NE England	1961–70	86.4	6.0	7.0	AC	Ditto
North Shields	NE England	1970–80	81.8	5.5	5.0	AC	Ditto
North Shields	NE England	1981–90	72.0	5.6	3.1	AC	Ditto
Marsden	NE England	1991–96	86.8	2.7	7.1	AC	Ditto
Marsden	NE England	1997–98	52		(1.5)	AC	Ditto
Skomer	S Wales	1980–90	86.6	6.7	7.0	CR	C.M. Perrins pers. comm.
Skomer	S Wales	1991–98	86.6	10	7.0	CR	Ditto
Skomer	S Wales	1999–2007	80.9	8.5	4.7	CR	Ditto
Brittany	France	1980–94	79.5	6.5	4.4	CR	Cam *et al.* 1998
Foula	N Scotland	1986–97	80.0	13	4.5	CR	Oro & Furness 2002
Isle of May	SE Scotland	1986–97 1986–2000	88.2 89.0	5.5	8.0 8.6	CR CR	Harris *et al.* 2000 Frederiksen *et al.* 2004b
Fair Isle	N Scotland	1986–97	83.0	12	5.4	CR	Rothery *et al.* 2002
Hornøya	N Norway	1989–2003	88.0	10	7.8	CR	Sandvik *et al.* 2005
Prince William Sound	Alaska	1991–95	92.0	?	12.0	AC	Golet *et al.* 1998
Middleton Island	Alaska	1987–90	93	2.4	13.8	AC	Hatch *et al.* 1993b

* *AC = accurate direct sightings; CR = capture–recapture estimates.*

THE RELATIONSHIP BETWEEN SURVIVAL RATES AND THE EXPECTED NUMBER OF BREEDING YEARS

While survival rates give a rough indication of how long individuals will live, a better indication is obtained from the expectation of (adult) life. If the survival rate of adults remains constant in relation to age, then it can be used to calculate the average expectation of life from the first year of breeding and the relationship between the two values is acutely curvilinear as shown in Figure 10.2. The number of breeding years increased rapidly for survival rates

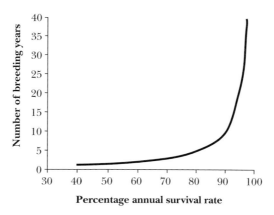

FIG. 10.2   *The relationship between the annual survival rates of adult birds and the number of years for which they can expect to breed. This assumes a constant survival rate with age. Note the rapid increase in breeding expectation at high survival rates. The survival rates of adult kittiwakes are in the zone 80% to 92%.*

above 80%. As a result, a relatively small increase or decrease in the survival rate of adult kittiwakes greatly influences the average lifespan and therefore the lifetime productivity of individuals. An increase in the annual survival rate of kittiwakes from 80% to 90% increases the number of breeding years from 4.5 to 9.5 years and a further increase in the survival rate to 93% increased the average number of breeding seasons to almost 14 years. As a result, small changes in the survival rate of kittiwakes have a pronounced effect on the average length of life and the number of breeding attempts each individual can make (Table 10.1). The potential effect on lifetime productivity of the high survival rate in Alaska is large, but is negated by the poor breeding success in most years.

These differences in survival rates indicate large differences in the expectation of life following breeding for the first time. At North Shields and several other colonies, kittiwakes bred on average for five or fewer years, for over eight years on the Isle May, and the Alaska birds bred for 12–14 years. It is amazing that in all of these colonies, kittiwakes continue to persist. The answer lies in recruitment of new breeding birds.

SEX-RELATED SURVIVAL

Consistently throughout the studies in the Tyne area of north-east England, males showed a lower survival rate than females and the average difference was three percentage points. The difference between sexes did not appear to vary with the overall survival rate. As a result of the differential survival rates, males on average bred for 4.6 years and females for 5.6 years. This difference during the study resulted in 56 females but only 32 males which bred for over ten years. One effect of this was that many of the oldest females had to mate with much younger males.

Most of the other studies did not examine the survival rates separately for the sexes, but Cam *et al.* (1998), Oro & Furness (2002) and Hatch *et al.* (1993b) failed to detect a meaningful effect of sex on the survival rates in their shorter studies, yet the difference persisted in every one of the averages for four-year periods at North Shields (Figure 10.3) and was obviously a regular effect over many years. Was this differential survival of the sexes unique to North Shields and if so, why? Will more extensive studies elsewhere eventually reveal sex-related survival?

AGE-RELATED SURVIVAL

In earlier analyses of the North Shields data, Aebischer & Coulson (1990) showed that the annual adult survival rate remained constant for the first 12 years of breeding, but then declined significantly in the group breeding for 13–19 years, with the mortality rate almost doubling and affecting the final 10% of the breeding birds.

In a study on the Isle of May, Frederiksen *et al.* (2004b) showed a similar increase in mortality when kittiwakes were 15 years old (i.e. they had bred about 11 times). However, the claim that the survival rate increased from 86% at two years of age to 93% at age 7–10 was not justified by their data. It arose because they made the assumption that a simple quadratic equation described the variation with age in survival in the kittiwake, and this, by chance, reached a maximum value at about ten years of age, with lower values in younger and older birds (see below for survival in relation to age).

In both studies summarised above, the decline found in the annual survival

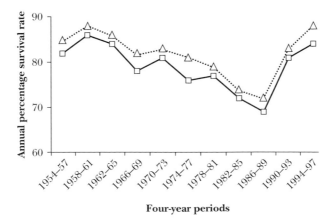

Four-year periods

FIG. 10.3 *Comparison of the survival rates of male and female kittiwakes in the Tyne area of north-east England in four-year periods from 1954 to 1997. Males are indicated by squares and the continuous line and females by triangles and the dotted line.*

rate of the oldest birds was reasonably attributed to senescence. However, it is necessary to be cautious about this interpretation because in both analyses, survival of the young birds was measured over a different range of years, with many years elapsing before any of the marked individuals became the oldest birds and their survival rates could be measured. In both studies, the overall survival rates of the kittiwakes decreased with year. Although an allowance for this was made by introducing a correction for year, it is possible that the correction was not sufficient.

To examine this possibility, and the question of whether survival rates varied with age, I analysed survival rates for male and female kittiwakes of all ages which were alive and exposed to the same risks of mortality in two five-year periods: 1973 to 1978 and 1979 to 1984. During these periods, some adult kittiwakes first ringed in the 1950s were still alive and had reached an appreciable age and kittiwakes over the whole age range were equally at risk. As a result, this analysis allowed a strict comparison of the mortality risks of birds of different ages in relation to the same five-year periods. In so doing, I avoided possible problems arising from comparing the survival rates of birds of different ages but in different numbers of years. Separate analyses were carried out for males and females.

The overall survival rates were appreciably different for the two time periods. Within each five-year category, there were negligible and no consistent changes in the survival rates in relation to age, and none showed significant trends (Table 10.2). Despite being based on large samples, there was little indication either that survival rates were different between the groups, or that the survival rate decreased in the oldest age groups. The same data sets, but also including birds over 17 breeding years, were also compared by survival curves which confirmed that up to 17 breeding years there was no indication that the survival rate varied with age in females (Figure 10.4) or in males. There is no justification is attributing different survival rates to young adults. Nor is there evidence from survival data of senescence existing in old kittiwakes; if it does occur, it can, at best, only affect 1–2% of the individuals which start to breed.

TABLE 10.2 *The annual mortality rates at North Shields of adult male and female kittiwakes in four age classes, all at risk to mortality between 1973–78 and 1979–84. Note that the survival rates in the oldest age classes in both sexes were very similar to the overall survival rates for all age classes combined and there was no indication of lower survival in the youngest or oldest age categories.*

	Males 1973–78		Females 1973–78		Males 1979–84		Females 1979–84	
Age	*N*	*Mortality rate*	*N*	*Mortality rate*	*N*	*Mortality rate*	*N*	*Mortality rate*
1–2	154	24.7%±3.5%	134	19.4%±3.4%	238	33.6%±3.1%	211	29.9%±3.2%
3–6	155	22.6%±3.4%	157	17.8%±3.1%	162	36.4%±3.8%	135	37.7%±4.2%
7–10	71	25.4%±5.1%	120	16.8%±3.4%	37	29.7%±7.5%	44	27.2%±6.7%
11–17	33	20.8%±6.9%	46	17.4%±5.5%	27	33.3%±9.0%	59	33.4%±6.1%
Total	413	23.7%±2.1%	437	18.8%±1.9%	454	33.5%±2.2%	449	32.5%±2.2%

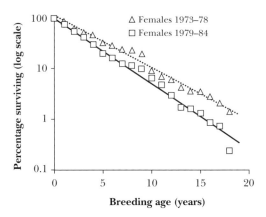

FIG. 10.4 *The age-related survivorship curves for female kittiwakes at North Shields during 1973–78 and 1979–84 (based on data used in Table 10.2) to examine the pattern of survival when birds of different ages were at risk to mortality over the same time periods. The linear slope of the logarithmic plot indicates a constant rate of survivorship, irrespective of age, and it was not until females had bred for more than 17 years that there was even a suggestion of decreased survival by which time senescence would only affect less than 1% of the females between 1973– 84. Comparable data for males led to the same conclusions.*

This conclusion would also be compatible with the appreciably longer life-span and expectation of life found by Hatch *et al.* (1993b) in the colony on Middleton Island in Alaska, where many more adult kittiwakes survived for more than 20 years.

Survival rates of immature kittiwakes

It is difficult to measure the survival rates of immature kittiwakes following fledging, because these young birds are oceanic for most of the time and the few ringing recoveries do not give a reasonably accurate measure of numbers dying, but rather the relatively few which were killed near human habitation. Probably the best estimate that can be achieved is to calculate the proportion of young fledged that reach breeding age.

If we assume that 100 pairs of kittiwakes fledge one chick per year and the population remains constant, then the survivors of the young equal the number of adults dying in a year. Taking the adult annual survival rate as 82%, then of the 100 pairs, 36 adults die each year and so 36 young are required to survive to breed to maintain the 100 pairs. This means that the survival from fledging to breeding would be 36%. If kittiwakes are assumed to breed on average at four years of age, this gives an estimate of 77.5% survival each year until maturity was reached when four years old. This calculation can be taken a little further by assuming that, as in most birds, the first-year survival is lower than in later years, and that the annual survival in these later years is the same as in breeding adults. In that case, the first-year survival rate would be 63%, that is, the risk of mortality in the first year of life is about twice that of adults, something which would appear to be acceptable as an estimate since in that first year, the young bird has to develop its own feeding skills.

Obviously, these calculations would vary with different productivity and adult survival figures, and for a population that is either increasing or declining. Delaying first-time breeding until age 5 results in the first-year survival rate needing to be increased to that of the adults – an unlikely outcome.

CAUSES OF MORTALITY IN KITTIWAKES

Predation by mammals on kittiwakes is extremely rare. Very occasionally, adults resting on the water have been taken by Grey Seals and a few individuals have been taken by Arctic Foxes while collecting nesting material. Both their cliff nesting sites and their oceanic winter habitat exclude most chances of mammalian predation. Predation by avian predators is relatively low in most areas. In the wintering area, such predation is unlikely since there are few predators in the open oceans. In the breeding season, adults are taken by falcons near some, but by no means all, colonies. Peregrine Falcons and Gyr Falcons in particular hunt kittiwakes over the sea near some colonies. Great Skuas appear to specialise in taking young and adult kittiwakes from nest sites on Foula (Shetland Islands), but the habit is not extensive (as yet) elsewhere. In some colonies in northern regions, White-tailed Eagles have recently become an increasing problem, taking kittiwakes from cliff sites particularly where the cliffs are not vertical, such as some sites in northern Norway and Russia. The role of predation on kittiwakes by large fish in the wintering area is unknown. Elsewhere, Glaucous Gulls and Ravens have been implicated in local predation on the contents of kittiwake nests.

From this review, it is evident that most kittiwakes die from causes other than predation. In a recent study, Coulson & Fairweather (2001) showed that the last breeding performance by kittiwakes was frequently less successful than in their earlier years. This is not merely the effect of senescence, since this 'last year' effect was also evident in young breeding kittiwakes where senescence is most unlikely. Of particular significance was the finding that the final breeding attempt was even poorer when it preceded the death of both partners in the following 12 months. These observations were interpreted as indicating that the poor breeding performance was caused by a deterioration in the condition of one or both of the parents some weeks or months before they died and involved a progressive deterioration over an appreciable period of time and which ultimately resulted in death. It is not known which diseases were involved in these cases, but avian tuberculosis (caused by *Myobacterium avium*) is a likely candidate and is widespread in birds. Swine erysipelas (*Erysipelothrix rhusiopathi*) and pseudotuberculosis (*Yersinia pseudotuberculosis*) have been isolated from gulls and the former was identified as the cause of death of one kittiwake at North Shields. The *uuku-virus* (a Bunyavirus) was isolated from a sick kittiwake near this colony (Eley & Nuttall 1984). The spirochete responsible for Lyme disease, *Borrelia burgdor-feri*, has been isolated from kittiwakes, as has flu virus, but again the effects of these on kittiwake health are not documented. Toxins produced by algal blooms can cause appreciable mortality in kittiwakes. An unidentified toxin caused the death of some 13,000 kittiwakes over two years in north-east England in the late 1990s. Starvation in winter, which appears to result in individuals being driven inland during storms, has produced numerous ringing recoveries well inland in Europe and North America, but the extent to which these contribute to the total annual mortality of kittiwakes remains unknown. Fortunately, mortality from oiling is rarely reported. There is much yet to be learnt about the causes of mortality in kittiwakes. However, currently, neither senescence nor predation appear to be important factors.

CHAPTER 11

Colonies, recruitment and population trends

COLONIAL breeding in the kittiwake means that the population is made up of colonies, each composed of many individuals. The additional category of colonies, which is absent in the populations of many birds, means that an additional level of abundance is recordable and available for study. Colonies have many characteristics; they have distinct patterns of growth, they attract recruits which replace the annual loss of breeding birds through mortality, and variation from the numbers required for replacement cause colonies to increase or decline in size. The summed sizes of all colonies measure the national and international abundance of kittiwakes and, over time, indicate whether numbers are increasing or decreasing. In this chapter, I examine the nature of colonies and how they grow, how new birds are recruited to colonies and, together with the previous chapter on survival rates which influence the numbers of recruits required in each colony, how the effects of recruitment and survival can be evaluated by repeated census work over periods of years.

THE NATURE OF KITTIWAKE COLONIES

Kittiwakes are obligate colonial breeders and colonies range in size from a few pairs (usually new colonies) to those comprised of over 100,000 pairs. The average size of a colony in England and in Norway is only about 300 pairs, and the majority range from about 100 to 1,000 pairs. Some colonies contain only kittiwakes, but in other places the narrow cliff ledges are shared with guillemots, Razorbills and Fulmars, while the nearby broader ledges are used by Shags and Gannets. In the Pribilof Islands, Alaska, Red-legged Kittiwakes always nest in mixed colonies with Black-legged Kittiwakes, and the two species show little difference in their nest site preferences.

WHAT IS A KITTIWAKE COLONY?

There is no simple or single answer to this question, and it seems that kittiwakes and humans see them as being somewhat different. Individual pairs of kittiwakes mainly regard their immediate neighbours as the 'important' group, and I can find no evidence that they in any way respond to the total size of the colony. On the other hand, many people have a general concept of colonies being discrete units, usually with continuity or near continuity.

Studies of how kittiwakes respond to each other during mutual social stimulation and courtship provide clues as to how kittiwakes see the interrelationship between pairs in the colony. As shown by Coulson & Dixon (1979), breeding kittiwakes appear to respond only to their immediate neighbours when engaging in mutual courtship displays, that is, those other pairs within a quite short radius. The effect decreases rapidly with distance and probably disappears within a radius of 3–5 metres. Pairs further away do not appear to respond or contribute to this social stimulation induced by neighbouring pairs for most of the breeding season. So Darling (1938), in expressing his ideas about social stimulation, was almost right but erred by suggesting colony size, rather than density of birds, played the important role.

If this concept is taken further, it leads to the suggestion that, from a behavioural perspective, a colony is a series of small interlinking groups. One pair

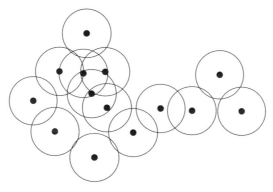

FIG. 11.1   *The position of kittiwake nests (solid dots) near the edge of a colony. The circles represent a 3m radius around each nest which indicates the approximate limits of interactions between displaying pairs and show the overlapping 'chain mail' type of effect.*

link with their immediate neighbours and these in turn link with other pairs and so on. Thus a colony seems to be like chain mail and is made up of a series of interlinking units, which together form an entity. An example of this is shown in Figure 11.1.

Early in the return and when vacating the colony at the end of the breeding season, the 'panic' flights involve the co-ordination of birds over larger areas engaging in synchronous departures from the cliffs, and may involve birds from over 200m of cliff, but these responses last only a few days and are not evident during the rest of the breeding season. The stimulus is visual, and individuals which are spread around headlands, so that some are out of sight of each other, do not usually display this synchronised response.

To ornithologists, the concept of a colony is more vague. Having encountered this, I used the term 'breeding station' in the original British kittiwake enquiry in 1959, but this term has not been used subsequently. The problem of whether a large aggregation of nests in a given area is one colony or several colonies reappears at each census, but a definition is not prepared and it is evident that no single definition is satisfactory. How much distance between two groups is necessary for there to be two colonies? Often a gap in the kittiwakes on a cliff is formed by a change in the cliff structure, for example by a change from a sheer cliff to a gradually sloping one without kittiwakes and then returning to a sheer cliff with kittiwakes again. Sometimes the gap is between small, nearby islands.

Consider the Farne Islands, off the Northumberland coast of north-east England. Five of the outer islands contain breeding kittiwakes, but with about 100–250m between neighbouring islands. Then there is a gap of about 2km to the inner group of islands, where there are three with nesting kittiwakes, again with 100–250m between them. About 3km from the inner island group, there is a also kittiwake colony nearby on the mainland coast.

So how many colonies are present? I arbitrarily divided the outer and inner group of islands into two colonies. Others grouped all the islands as one colony, but when the colony on the mainland was established, it was immediately considered separate. Elsewhere, there are cases where kittiwakes nest on opposite sides of islands and obviously never interact there, but these are usually referred to as a single colony.

Over the years, along the north-east coast of England, I have seen distinctly separate groups of nesting kittiwakes increase in numbers and spread along the cliffs towards each other, where they eventually coalesce. Such amalgamations have probably occurred when forming the very large colonies that currently have tens or even hundreds of thousands of pairs.

For convenience, I separate groups of breeding kittiwakes which are more than one kilometre apart, but this is entirely arbitrary. Even the addition of 'sub-colonies' does not help a great deal because it also requires an arbitrary definition. Perhaps kittiwakes know best about what is a functional group, but they are not good at making a census!

Typically, kittiwakes nest on high, vertical cliffs, usually with the sea reaching the base, at least at high tide. In a few places in Greenland, kittiwakes nest on inland cliffs, a kilometre or more from the sea, but these are exceptional sites. Extreme inland nesting sites occur along the River Tyne at Newcastle and Gateshead in north-east England, almost 18km from the North Sea.

On very high cliffs over 100m high, kittiwakes almost invariably nest on the lower half and often even nearer the bottom. On lower coastal cliffs, they nest over the entire height of the cliff, apart from avoiding a narrow band at the top that can be reached by predators and the bottom of the cliff, where rough seas would dislodge nests and their contents. On some small islands, presumably because they lack mammalian predators, kittiwakes nest on steep sloping cliffs, as on parts of the Farne Islands in Northumberland and on some islands off the northern coast of Norway. However, this type of site is rare on mainland coasts, presumably because the nests can be reached by mammals. At some sites, kittiwakes nest within the mouths of caves and in other places they build nests on sites that have overhanging rocks above which provide shelter from rain, but I do not believe these sites are specifically selected by the kittiwakes for protection.

In Spain and Portugal, the few breeding kittiwakes only nest on north-facing cliffs and where the sun does not shine directly on the nests. Presumably this selection is deliberate, and made to avoid the intense heat from the sun, which would probably greatly reduce the breeding success. It seems likely that the adverse effects of intense solar radiation on breeding kittiwakes may well define the southern geographical breeding limits of kittiwakes.

Kittiwake nests are invariably built on narrow ledges, but they need a minimum area of about 80 by 80mm of near-level surface upon which to construct a nest. Sometimes kittiwakes use wider ledges, but more so where auks and Fulmars are absent, since both species often competitively exclude kittiwakes from such sites. An exception to this exclusion was recorded at Marsden in north-east England, where kittiwakes in an expanding colony took over sites that had been used previously only by Fulmars. The kittiwakes took advantage of the periodic pre-breeding departures of Fulmars from the sites to occupy them and establish ownership, and attacked the returning Fulmars in mid-air as they approached the cliff. If the Fulmars succeeded in landing, they were then victorious, helped by ejecting oil from their beaks toward the kittiwakes, but in the long run, kittiwakes took over the ledges (Coulson & Horobin 1972).

As the numbers of kittiwakes increased in parts of the North Atlantic during the 20th century (there is little historical data from the Pacific), many new colonies were established and the majority of these were on lower cliffs. This trend was probably caused by most of the higher sea cliffs already supporting kittiwake colonies and also by the reduction in shooting and egg collecting and the introduction of legal protection that allowed kittiwakes to nest without persecution on low cliffs.

The use of man-made structures as nesting sites for kittiwakes began in 1931 near Edinburgh, Scotland, where kittiwakes nested on a pier. Subsequent nesting on buildings occurred at Dunbar (south-east Scotland) and then at North Shields, where I based most of my study. Nearby, many buildings were used up to 21km inland from the coast along the River Tyne, and similar sites also occurred elsewhere on the east coast of England, in Norway and briefly in Denmark. The use of buildings is considered in more detail in Chapter 13.

Kittiwakes have nested on many other structures. In Alaska, they have nested on the superstructure of a wrecked ship and also on ledges of a tower on Middleton Island, where Scott Hatch has centred his studies. Around Britain, they nest on offshore gas rigs in both the North Sea and the Irish Sea, and also on top of the wooden support pillars of a pier at Hartlepool, which is used to transfer seawater to a nearby factory. Near North Shields, the kitti-wakes nested for a number of years on a gantry at a ferry terminal before it was demolished. The trend toward using lower 'cliffs' reached its extreme when kittiwakes nested on a storm beach on the island of Hirsholmene, off Frederikhavn in Denmark. The nests were placed on and against boulders up to a metre in diameter and these kittiwakes later spread onto the adjacent low sand dunes. At about the same time, two pairs of kittiwakes nested among Sandwich Terns on a low sand dune at Scolt Head in Norfolk, England.

NEST SITES

The kittiwake's nest is bonded to the rock or other surface, and is a drum-shaped, somewhat untidy structure about 300mm in diameter which often overhangs the limits of the ledge. Kittiwakes defecate over the edge of the nest and while much of this falls into the sea, strong winds cause the faeces to drift onto the sides of the nest and the cliff faces below. As a result, the cliffs receive extensive 'whitewashing', making the presence of a colony evident, even from a distance.

Such nest-side defecation can cause a considerable disadvantage to kitti-wakes nesting lower on the cliff face, particularly for those nesting on sloping rather than vertical cliff faces. While adults frequently bathe and can remove the 'whitewash' which sometimes rains down upon them, the chicks in such nests remain on the nest for over five weeks and can become extensively soiled and stained on these less favourable sites. Despite preening, the soiling may lead to the feathers losing their waterproof qualities, resulting in the chicks looking particularly bedraggled during and after heavy rain. Soiling probably appreciably reduces their chances of surviving after fledging and landing on the sea, but this has yet to be proven. There is probably an advan-tage in nesting high within the colony, so reducing the problem produced by receiving 'gifts from on high'.

Colony growth

Surprisingly, little has been published on colony structure and growth for seabirds and for the kittiwake in particular. New colonies of kittiwakes are usually formed by between three and twenty nesting pairs, with additional non-breeding individuals usually present. The infrequent exception to this pattern is when an existing colony becomes unsuitable, such as results from major disturbance or adverse conditions (as has occurred in France, Cornwall, North Shields and the Isle of Man), and many experienced breeders, which otherwise would be colony faithful, relocate to a new site.

During the 20th century, many new colonies were formed, and some of these have had the numbers of nests recorded in several years, so that the pattern of growth is well known. Between 1959 and 1969, when kittiwake colonies in the British Isles were all increasing in size, I was able to obtain information on the increase in size of 46 colonies during the same ten-year interval. This was possible through a national census I organised with the British Trust for Ornithology. The results have been converted to annual (compound) rates of increase and plotted in relation to the size of each colony in 1959 (Figure 11.2). These results gave the first indication that small (young) colonies increase at a very high rate, while large (old) colonies had much smaller annual rates of increase.

The collection and accumulation of other annual data on colony sizes produced further information. For example, the detailed examination of the size of the colonies at Marsden, in north-east England, showed that after the first few years of breeding there (unfortunately the numbers of nests were not counted in these years), they increased by a similar number of nests (about

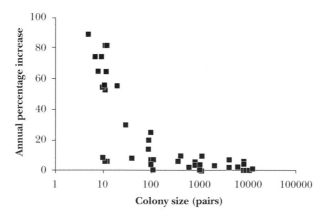

FIG. 11.2   *The annual rate of increase in size of 46 kittiwake colonies in the UK between 1959 and 1969 in relation to colony size in 1959. Based on the British Trust for Ornithology national counts in these years (Coulson 1963a, 1974). Note the much higher rate of increase in colonies which were small (and newly formed) in 1959, compared with the larger colonies. Colony size is presented on a log-scale.*

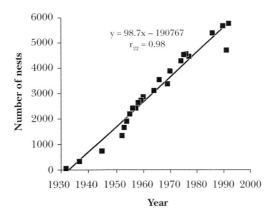

FIG. 11.3   *The number of kittiwake nests at Marsden, Tyne and Wear, north-east England from 1932 to 1995. The colony was first formed between 1929 and 1931. Apart from a deviation between 1937 and 1953 (possibly an effect of World War II?), the numbers have increased linearly at about 100 extra nests each year throughout, and numbers have not followed the exponential curve which would have been expected in the growth of an unrestrained population (see text). Note the record of 7,700 nests at Marsden in 1986 (Brown & Grice 2005, p. 386) is erroneous.*

100) each year (Figure 11.3) for the following 60 years. Growth did not occur in the exponential manner that would be expected if it followed that of an unrestricted population, despite more and more individuals becoming available to produce young.

The difference between an exponential (uncontrolled) increase in numbers of any organism and one that is progressively restrained is best demonstrated graphically. Instead of plotting the numbers of pairs against time (year) as in the hypothetical example used in Figure 11.4A, it is much more informative to first convert the numbers of pairs to logarithms and plot these against time. If this is done when an exponential relationship exists (Figure 11.4A), the trend line becomes linear (Figure 11.4B), and the slope of the line is informative because it is a direct measure of the rate of increase. As a result, it is then easily seen if and when the slope of the trend line changed, that is, if and when the rate of increase changed. So plotting the same hypothetical data used in Figure 11.4A on a logarithmic (base$_{10}$) plot in Figure 11.4B readily reveals that the rate of increase remained unchanged over time and that the growth was unrestrained. By applying this transformation to the Marsden data (Figure 11.5), it is evident that the slope of the graph changes progressively with time, implying that the growth was being restricted more and more as the numbers increased.

Over the years, I have collected the numbers of nests in a series of colonies over a series of years. Examination of the rates of increase of kittiwake colonies with time almost always showed the same pattern. On a logarithmic plot, the relationship was usually a curve, showing decreasing rates

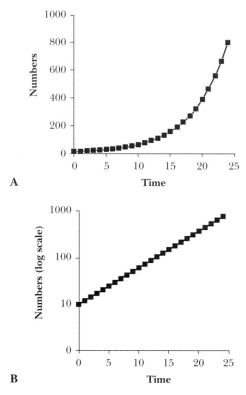

FIG. 11.4   *The effect of plotting an exponential growth curve (A) on a logarithmic plot (B). The constant slope of the line on the logarithmic plot shows that the rate of increase did not change over time.*

of increase as the colony grew. Examples of these curves are given in Figures 11.5, 11.6 and 11.7 for three colonies, to indicate the general nature of this effect.

It is possible that the reduced rate of increase as a colony grew and became older was a result of increasingly adverse conditions affecting the survival and subsequent recruitment of young kittiwakes. The possible influence of environmental changes over time is excluded in Figure 11.8, where the growth for old and new colonies was measured in the same years. It is evident that the slope of the trend line is appreciably and consistently steeper in the new colonies than in the older, larger kittiwake colonies at Marsden during the same period. As colonies grew, the percentage rate of increase became less. So why do small colonies increase more rapidly? In relation to their size, young colonies must be more attractive to recruits, but why is this so?

A clue to the answers for these questions is revealed when a single colony is divided into several horizontal but arbitrary sections, from the centre to the edges, and the growth measured in each. Figure 11.9 clearly shows that the rate of growth between the centre and the sub-centre differed and this differ-

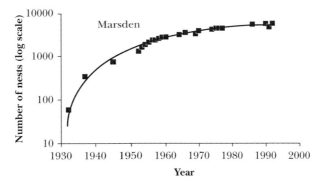

FIG. 11.5 *The numbers of kittiwake nests at Marsden, Tyne and Wear, north-east England, 1932–1995 plotted on a logarithmic scale against year. The resulting curve shows that the rate of increase progressively decreased as the numbers of kittiwakes increased. A best-fit polynomial has been fitted to the data.*

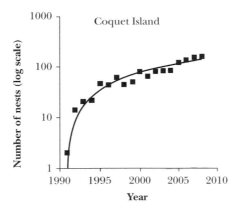

FIG. 11.6 *The numbers of kittiwake nests on Coquet Island, Northumberland, England, plotted on a logarithmic scale against year. The first breeding was in 1991. A best-fit polynomial has been fitted to the data.*

ence increased even further at the edge of the colony. Over the 14 years when the sections of this colony were counted, the number of nests in the centre increased by only 6%, whereas the sub-centre increased by 67% and the edge by a massive 375%. In reality, during the 14 years, most of the sub-centre acquired the characteristics of the centre in that it became fully occupied, while some of the original edge became similar to the original sub-centre. The colony increased from 198 to 388 nests in 14 years, but in considering where new birds nested, 70% of the additional nests occurred in what was originally the edge and the immediate areas beyond this, while only 9% took place in what was perceived as the centre of the colony. As a colony grew, the extent of centre and sub-centre became larger but the edges tended to remain the same size. With most of the increase restricted to near the edge, the

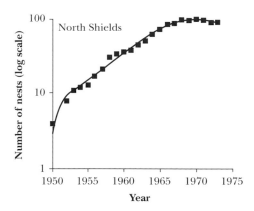

FIG. 11.7   *The growth of the kittiwake colony at North Shields from 1950 to 1975, after which time numbers of nests tended to remain at the same size.*

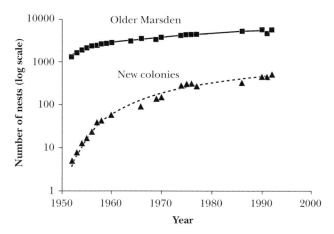

FIG. 11.8   *A comparison of the rate of increase in numbers of kittiwakes nesting at the older, larger colonies at Marsden and in two new colonies established in 1953 and 1954 (average number of nests at the new colonies presented). Note that throughout, the slopes of the curves for the new colonies in the same time periods are steeper, indicating higher rates of growth than at the older Marsden colonies, although by 1990 the lines were becoming almost parallel.*

overall rate of increase of the colony as a whole declined because the edge became a smaller proportion of the whole colony.

Similar conclusions were reached after mapping the growth over time of the original area colonised by kittiwakes on Coquet Island, Northumberland. Breeding started in 1991, and nest positions were mapped in 1992, 1995 and 2000 (Figure 11.10). The interpretation of the growth is presented in Table 11.1, and a number of conclusions can be drawn from these data.

1.   The growth of the colony centred around the area first occupied, that is, new recruits were attracted to the original group.

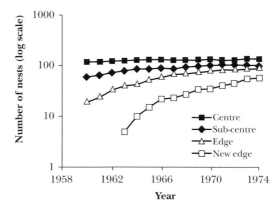

FIG. 11.9   *The change in number of nests in four areas of a colony at Marsden, 1958–1974. In 1963, the colony had spread beyond the previous edge area, forming a new edge area. Note how the rate of increase (slope of the trend lines) varied within areas of the same colony. The sub-centre and edge areas were on both sides of the centre area. Between 1960 and 1974, the centre increased by 14%, the sub-centre by 67% and the edge by 375%.*

2.   Despite increasing from 10 to 51 nests, the length of occupied cliff increased less rapidly because the density of nests progressively and appreciably increased.

3.   There were relatively small increases in the number of pairs nesting beyond the colony limit over the time intervals considered. Between 1992 and 1995, only seven of twenty new nest sites contributed to increasing the length of the colony, while between 1995 and 2000 only three nest sites extended the colony limits.

4.   There was an increase of nine new nest sites in the centre of the colony between 1992 and 1995, but only four between 1995 and 2000, because most of the possible sites where nests could be built were already occupied.

5.   Most growth of the colony occurred in the sub-centre areas, between the central area and the previous edge of the colony, with 11 new sites (55% of the new nest sites) between 1992 and 1995 and 14 new sites (67% of new sites) between 1995 and 2000.

6.   The density of nests increased over time (Figure 11.11 and Table 11.1).

These figures suggest that there are two factors which restricted where the kittiwake colony grew. First, fewer pairs managed to become established in the centre of the colony because virtually all of the suitable sites were occupied by 1995. Second, recruits were reluctant (or unable) to establish nest sites far beyond the existing limits of breeding pairs, and so the limits of the colony increased only slowly.

The main reason for the progressive differences in growth is that a kittiwake colony is not an isolated population, but depends upon the influx of

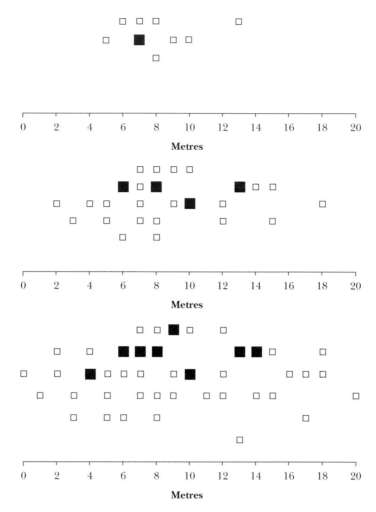

FIG. 11.10   *The position of kittiwake nests in the original colony on Coquet Island plotted on a 0.5m (height) by 1m (length) grid. Each open square represents a single nest within a grid cell and filled squares are pairs of nests within a grid square which are virtually touching each other. Top: 10 nests in 1992. Middle: 30 nests in 1995. Bottom: 51 nests in 2000.*

recruits from many other colonies, with some recruits coming from hundreds of kilometres away.

These conclusions should not be interpreted as indicating that kittiwake recruits avoid the centre of the colony. On the contrary, there is strong demand to obtain sites there, but there is soon a shortage of unoccupied ledges, and recruits entering the centre of the colony do so mainly by replacing adults which had bred there and had died since the previous breeding season.

In effect, colonial organisation acts to limit the numbers in the colony

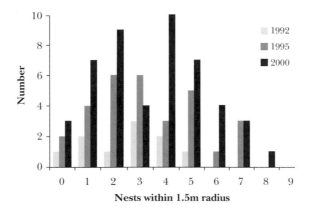

FIG. 11.11 *Comparison of the density of kittiwake nests on Coquet Island in 1992 (second year of breeding), 1995 and 2000. Density was measured by recording the number of other nests within a radius of 1.5m around each nest. Note how the average density and the spread of densities increased as the colony grew (less edge effect), but that there were always nests without another nest with 1.5m and these were usually near the edges of the colony.*

through competition for the finite number of nesting sites within or at the edge of the colony. As the colony becomes larger, the proportion of available sites becomes fewer. This may be seen as an example of what is more frequently referred to as a density-dependent effect which regulates colony size. However, it does not prevent new colonies being formed and therefore does not necessarily regulate the size of the population. This would only occur if the proximity of other colonies influenced colony size as suggested by Furness and Birkhead (1984).

While recruits are required to increase the size of a colony, these are in addition to those needed to replace the mortality of adults which bred there in the previous year. Obviously, the numbers needed to replace the mortality increase with the size of the colony. This can be investigated further by returning to the increase in numbers of nests at Marsden, where the number of additional pairs was essentially constant over time. By taking a realistic mortality rate of 15% per year for adult kittiwakes and applying

TABLE 11.1 *The characteristics of the new kittiwake colony on Coquet Island, Northumberland in the early years of its growth. Comparisons are made with the distribution of nests in 1992, 1995 and 2000.*

Year	Size (nests)	Length of colony	Mean number of nests within 1.5m radius	New nests beyond earlier limit of colony	Additional nests in centre	New nests between centre and previous edge
1992	10	8 m	2.6			
1995	30	16 m	3.3	7 out of 20	8	5
2000	51	21 m	3.8	3 out of 21	4	14

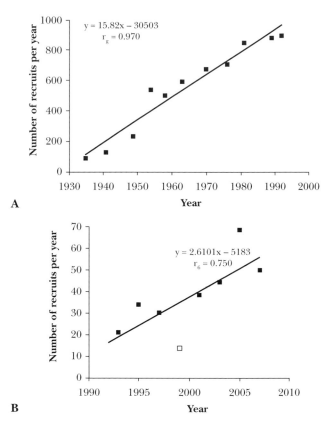

FIG. 11.12 **A.** *The estimated number of pairs of new kittiwakes breeding each year at Marsden, based on nest counts and an assumed mortality rate of 15% per year. Note that the numbers of recruits continued to increase over the first 60 years of the colony's existence, although the number of additional nests each year remained almost constant. There is a suggestion that the relationship shown may be slightly curvilinear but this is not significantly so, and at best it is a weak effect.* **B.** *The number of pairs of new breeders averaged for each of two years on Coquet Island, assuming an adult mortality rate of 15% per year. The point for 1998 and 1999 was a period where there was an exceptionally high mortality of adults in north-east England and therefore this point probably underestimates the actual recruitment in each of these years. However, it has been included in calculating the best-fit line.*

these to the numbers at Marsden, it is possible to estimate the total of new recruits joining the colony each year, i.e., causing the increase and replacing adult mortality. Figure 11.12A shows that despite the constant increase each year, the total of recruits increased progressively over time. Data obtained for the Coquet Island colony are for a shorter period of time (Figure 11.12B), but show the same trend; colonies need to attract more recruits as they grow in size. The proportion of recruits in these colonies which contribute to the growth has been estimated (Figure 11.13A and B). While the total numbers of recruits to the colonies increased annually and in a

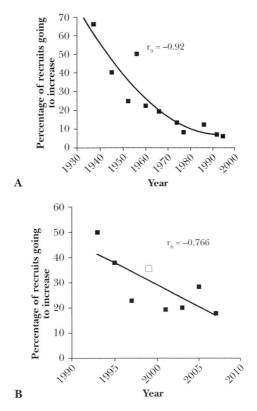

FIG. 11.13   **A.** *The estimated proportion of new breeding kittiwakes that contributed to the increase in size of the colony each year at Marsden from its first establishment. Based on the census of nests shown in Figure 11.3. Such a declining trend will tend to limit the size of the colony.* **B.** *Comparable data for Coquet Island from 1994 to 2007, during the early development of this colony, but each point represents a two-year period. The open square indicates a period when the adult mortality rate was probably higher than 15% per year and so overestimates the proportion which contributed to colony growth. This point has been included in determining the best-fit line, but if it was excluded, the relationship would then be significantly curvilinear.*

linear manner, the proportion of these which contributed to the increased number of nests and breeding pairs decreased progressively as the colonies increased in size, again demonstrating the regulatory nature of colony size and an effect which would bring about an equilibrium in the colony size, providing breeding productivity and adult mortality rates did not change.

Kittiwakes normally breed for the first time when four or five years old, and therefore all colony increase in the first few years must be due to immigration of birds reared in other colonies. In fact, it is many years before the return of young born in the colony (philopatry) becomes appreciable, and in detailed studies of recruitment (see Figure 9.11), the number of immigrants exceeded the number of philopatric individuals for many years and probably continues permanently.

The expansion of the edge area of a colony is restricted to a few metres beyond the limit already set by established breeders. This situation can be readily modelled, based on the following:

1. Recruits have a strong preference to occupy nest sites in the centre of the colony.
2. Availability of nest sites in the centre is primarily controlled by the death of previous site owners who have died. After the first few years of breeding, all suitable sites in the centre of the colony are occupied, so that recruitment there is restricted by the mortality rate of adults.
3. In the area between the centre and the edge, there are some sites which are still available, and others become available through the mortality of existing breeders.
4. There are many physically suitable sites at the edge of the colony, but only those in close proximity to established breeders are acceptable, because new breeders need the social stimulation of neighbouring pairs.

WHAT HAPPENS IN COLONIES THAT DECLINE IN SIZE?

During most of the years I have studied kittiwakes, their numbers have progressively increased. It was not until the sudden, high mortality of adults in the Tyne area of north-east England in 1998–1999 that the numbers nesting at Marsden declined markedly. I had previously considered, in theory, what would happen in a situation of declining numbers. I expected the expansion of the colony to cease and that recruits would first colonise available sites in the centre and sub-centre of the colony. Since I believe that the edge of the colony is the least favoured area for recruits, this area would cease to attract new breeding birds and existing birds breeding there might also move towards the centre of the colony. The effect would be that the extent of the colony would tend to shrink from the edges, but the high density areas in the centre would be maintained.

However, this did not happen in this simple manner at Marsden, primarily because I had overlooked the intensive site faithfulness of established breeders who maintained a high breeding success. As a result of a 65% decrease in the numbers of breeding birds over two years following the high mortality, the length of cliff used by breeding kittiwakes declined by only 5%. The main effect was that the density of nesting birds declined. Ten years after the mortality event, numbers nesting at Marsden had only increased by 9% and the overall density of nesting birds had not recovered to that of the early 1990s. New recruits joined near existing groups, and were attracted to what had been the central areas. As a result, some high density areas were recreated, but the main change was more extensive areas with low densities of nesting kittiwakes, with numerous potential nesting sites left vacant. The expected reduction in the limits of the colonies did not occur, nor did the immediate reforming of dense groups.

NEST DENSITY

The natural sea cliffs used by kittiwakes vary considerably in their structure and composition. On cliffs composed of hard, igneous rocks, the nature of the cliff remains little changed over time, and the same rock ledges are used year after year by nesting kittiwakes. The density of nesting kittiwakes on such cliffs is often low, as shown for Dunstanburgh, Northumberland (north-east England) in Figure 11.14.

Kittiwakes also nest on cliffs composed of sedimentary rocks, and many of these are much more susceptible to erosion. As a result, ledges are often more numerous (Marsden in Figure 11.14) and high densities can occur, but ledges available for nests in one decade may have disappeared by the next. This has been particularly evident at the Marsden colony in north-east England, where sea cliffs are composed of magnesian limestone (dolomite), and the rock strata vary considerably in hardness, with differential erosion producing ledges. These ledges are subject to further erosion, changing some flat ledges into steep slopes, and making them incapable of supporting kittiwake nests. It also appears that the activity of the kittiwakes locally increases the rate of erosion. Further, the cliff bases are subject to erosion by the sea, and cliff falls are not uncommon. Some of these cliff falls produce new nesting sites, while others remove suitable sites and replace them by sheer faces without suitable sites. As a result of erosion, two cliff faces at Marsden that contained over 500 and 270 nests each in the 1950s, and had the highest density of nesting kitti-

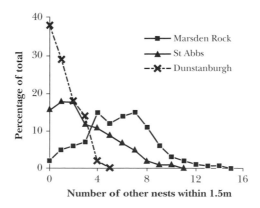

FIG. 11.14 *The variation in nest densities between kittiwake colonies, expressed as the percentage of nests which had the given number of other nests within a 1.5m radius of each nest in the colony.*

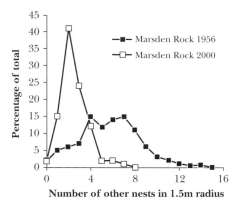

FIG. 11.15 *The change in density of kittiwake nests in the same part of the Marsden colony in 1956 and in 2000, measured by the number of other nests within a 1.5m radius of each nest in the colony. Most of this change over 44 years was caused by erosion of the ledges on the magnesian limestone cliff, which changed many horizontal ledges to steep gradients which were then unable to support kittiwake nests.*

wakes that I have seen, have progressively lost many of the nest ledges through erosion. By 2008, those areas supported only 102 and 76 nests respectively, and with few suitable nest sites remaining unoccupied. The change in nest density here has been appreciable (Figure 11.15) and the time of breeding here has become both later and more synchronised.

The density of nesting kittiwakes is determined through nest site selection by the birds, but the ultimate factor controlling density is the rock structure of the cliffs within the colony, which in turn is determined by regional geology.

So kittiwake colonies become saturated at the centre, the main expansion occurs between the centre and the edge of the colony, and the expansion of the limits of the colony is markedly restricted to the immediate edge of the colony and even then, relatively few kittiwakes start breeding there.

It has been thought by many ornithologists writing in the past century that a colony is an isolated and discrete population, and is entirely responsible for producing young to replace the natural adult mortality. This concept has often been applied to seabird colonies in general, and to the kittiwake in particular. But it is not true and is not even a reasonable approximation to the real situation. Coulson & Neve de Mevergnies (1992) showed for the first time that many young kittiwakes breed far from the place they were hatched as chicks. In a subsequent study, Coulson & Coulson (2008) produced evidence that only a minority of surviving young kittiwakes returned to breed in the colony in which they were born. In the past, it also had not been appreciated how difficult it was to locate all of the individuals that moved away from their natal colony to breed.

As a colony grows, most expansion extends laterally, but this results in the edge of the colony remaining much the same size as in a small colony, while the centre and sub-centre area, where most of the nest sites are already occupied, increases with the size of the colony. Thus the area where many of the recruits can obtain a site, namely, the zone within the edge, becomes proportionately less, and this makes the colony as a whole less attractive. Further, as a colony increases in size, it requires more recruits to replace the natural adult mortality. These two factors together are capable of producing the slow increase in size. A similar effect has been suggested for the slower growth with size in Gannet colonies (Moss *et al.* 2002).

TYPES OF NEST SITES

Colonies impose a major restriction as to where individuals can breed. Kittiwakes have to breed within the colony or at its immediate edge, and do not (and cannot) go elsewhere and breed as isolated pairs. Julie Porter made a detailed study on recruitment at the North Shields colony over three consecutive years as part of her Ph.D. study, and I rely much on her investigations in writing this chapter on recruitment.

Before breeding, the recruiting males must obtain a small ledge suitable to support a nest. Young birds hoping to breed arrive in the colony late in the season and long after experienced breeders have arrived and selected their breeding sites, which in many cases are the same as they used in previous years. As a result, there are a limited number of sites available to a potential first-time breeder, and these form three categories. First, there are some sites in the colony that have been used before, but the previous owners had died or moved site since the last breeding season and so these remained unoccupied. Such sites comprised between 10% and 20% of those in use the previous year. Second, there are some potential sites immediately adjacent to established pairs and their nests, but there is often a problem in using these due to the aggression of the established birds. Third, there are unused sites immediately around the edge of the colony, and the use of these causes the extent of the colony to expand. Potential sites at a greater distance beyond the edge of the colony were not used, and so the available area at the edge of the colony is quite restricted.

In a growing and expanding colony, all three types of sites are used by new recruits, but in a colony that is decreasing or stable in numbers, only the first two types take preference and the third is not used by recruits.

The central part of the colony is considerably attractive to recruits. This is the area of greater breeding success (than at the edge), and this preference means that should the breeding numbers decline, the colony structure will tend to be retained and the decrease will tend to occur near the edges. However, in the centre of the colony, most unoccupied sites are those left vacant through mortality of previous owners and there is considerable competition from potential recruits for any that become available. As a result, these vacant sites are rapidly occupied as the first choice among recruits. This intense attraction was readily demonstrated by temporarily removing the site owners present on a nest site in the centre and on one near the edge of the colony. The owners were removed for 30 minutes and this left both sites unoccupied. In that short period of time, one after another of six prospecting males landed on the centre nest site, were captured and were temporarily retained. In the same 30-minute period, no prospectors landed on the site near the edge of the colony. I repeated this simple experiment several times prior to egg laying, with very similar results, and only once did a prospector visit the site near the edge. Attracting recruits to the central sites also proved to be a very successful way of capturing prospectors for marking.

Some potential sites throughout the colony are difficult to occupy because of the aggression of the birds in possession of an immediately adjacent nest. Attempts to occupy such sites in the pre-laying period often resulted in fights in which the recruits were repelled. As the breeding season advanced, two factors eventually changed the situation.

First, once the resident birds on the adjacent site had laid and started incubation, their aggression was reduced, and while they continued to jab at a bird landing on the adjacent site, they did so without rising from the eggs and so had a much reduced reach. Thus once the established pair had laid, only then were recruits tolerated on the adjacent site. Consequently, this resulted in the recruits pairing and laying very late and often three weeks or longer after their established neighbours.

Second, on some sites the nest material from the previous year remained and this accumulated year on year. While the site owners in the past could bill jab at a recruit attempting to land alongside them on the same ledge, once their nest has reached a height of over 20cm, they could no longer reach a recruit landing alongside on the bare ledge. As a result, the recruit avoided attacks and a new site became occupied immediately adjacent to and on the same ledge as the established pair. Frequently, these methods increased the nest density and involved a pair with several years of experience breeding adjacent to a pair nesting for the first time and at very different stages in the breeding cycle.

It was evident that the established pairs soon recognised the new arrivals as individuals and ceased to threaten them. Once such tolerance was established, it was retained in future years and there was no longer aggression between the two site owners nesting side by side. Confirmation of tolerance became evident when I removed the taller of the two nests during the winter, making both nests of similar height. This did not result in a resurgence of aggression, nor the exclusion of one of the pairs, when all four adults returned for the next breeding season. However, the arrival of a stranger on one of the sites while the owner was absent was immediately greeted by intense aggression.

Vacant sites in the centre of the colony tended to be occupied first, followed by vacant sites near the edge of the colony, while the last to be occupied were the doubling-up sites which tended to be at the centre of the colony. Those recently recruiting at the edge of the colony tended to spend a smaller part of the day in occupation and sometimes the site was left unoccupied for hours. Had birds at the centre behaved in this way, rivals would have taken over the sites. Gaining a central site required one of the recruiting pair to be present there for most of the daylight hours to protect ownership, and so this becomes a major new commitment in time and effort, while at the edge of the colony, the investment is not so severe.

EVIDENCE THAT RECRUITS ARE YOUNG, FIRST-TIME BREEDERS

The new breeders arriving without rings at North Shields were young birds breeding for the first time. Few, if any, of these individuals had previous breeding experience in other areas and then had subsequently changed colonies. Evidence for this is as follows:

1. Convincing evidence comes from comparing the first breeding performance of birds which arrived at the study colony as unmarked individuals with those ringed as chicks and of known age (Table 11.2). Both categories of recruits showed similar poor breeding performance, including late laying compared with the colony as a whole, a low clutch size, and comparable but low productivity of young fledged per pair. Experienced birds showed much earlier breeding and higher breeding productivity.

2. The great majority of first-time breeders were present in the colony for one or more years before breeding. Typically, they arrived at the colony late in the season, similar to young birds of known age.

3. Many first-time breeders retained one or more dark-tipped feathers in the wing coverts and had more extensive black on the tip of the tenth primary in each wing, which occurs in a majority of birds known to be two-year-old birds (Coulson 1959). Many had yellow areas on the outside of the 'knees' when first captured, which turned black within a year or so. The orbital ring was often dark red rather than bright orange.

TABLE 11.2 *A strict comparison of the breeding performance of kittiwakes breeding for the first time which were ringed as chicks and of known age and those which were unringed. Based on 207 birds in each category. Each ringed bird was paired with the nearest unringed, first-time breeder nesting nearby, in the same year and of the same sex.*

	Ringed	*Unringed*
First egg of clutch	24.7 May ± 0.7 days	22.7 May ± 0.6 days
Clutch size	1.82	1.84
Breeding success	50% ± 3.4%	49% ± 3.5%
Productivity per pair	0.91	0.90

4. Birds ringed as chicks were mainly 2–4 years old when they first arrived in the colony.

5. Over 40 years, only two colour-ringed breeding adults from the North Shields colony moved to another colony, and in these cases the moves were associated with the trauma of their mates being shot at the colony. Adult kittiwakes are known to change colony due to major disturbance and also when breeding success is repeatedly low. At other colonies in the North Shields area, breeding success was consistently high, disturbance was minimal and numbers increased. Adults marked in other colonies in the area were not known to change colony. Based on this evidence, it can be presumed that few adults with previous breeding experience elsewhere moved into the study colony. This evidence, and the low degree of philopatry, strongly indicated that the immigrants of both sexes into the study colony were almost entirely young birds and that they had not bred elsewhere.

THE BREEDING PERFORMANCE
OF KITTIWAKES OF KNOWN AGE

A sample of males breeding for the first time when known to be three years, four years and older (Table 11.3) showed a progressively better performance when breeding was delayed to a greater age, but the differences were small. The data for females were based on smaller samples, but again there was only a hint of marginal improvement when breeding for the first time was delayed. The performances of all classes of first-time breeders were later and lower than those of adults who were known to have bred more than three times and were therefore older, but were of uncertain age.

The improvements in breeding performance of kittiwakes of known age between the first and second occasions were appreciable, and were comparable to the change in first breeding at different ages (Table 11.4). There were no differences between males and females and the data have been combined. While productivity improved by about 13% in males and by 5% in females breeding when less than five years old compared with those over four years old, the change in productivity between the first and second time of

TABLE 11.3 *The breeding performance of male and female kittiwakes breeding for the first time in relation to their age when so doing.*

	Male age 3	Male age 4	Male age over 4	Female age 3 or 4	Female age over 4
N	61	47	40	29	25
Date of laying (May)	24.6	23.4	22.9	22.3	21.4
Clutch size	1.77	1.80	2.00	1.84	2.07
Young fledged per pair	0.86	1.05	1.08	0.95	1.00
Breeding success	49%	58%	54%	52%	48%

TABLE 11.4 *First- and second-time breeding performance of male and female kittiwakes of known age and which bred for two or more years.*

	N	First breeding year	N	Second breeding year	Improvement
Date of laying (May)	165	25.2 ± 0.6	165	18.1±0.7	7.1 days earlier
Clutch	165	1.84	165	2.00	0.16 more
Fledging	165	0.95	165	1.23	0.28 more
Success	165	52 ± 4%	165	62 ± 4%	10% more
Total failure	165	39 ± 4%	165	22 ± 4%	17% less

breeding was 29%. A similar comparison of the percentage breeding success showed that those breeding for the second time improved by ten percentage points more, whereas in first-time breeders, differing by a year showed little or no improvement.

AGE AT FIRST RETURN AND BREEDING

The information on the age of first return and first breeding relates to those marked individuals which returned to breed at North Shields. In Julie Porter's intensive and detailed study, spread over three years on marked recruits to the North Shields colony, she found that every first-time breeder was seen in or on the edge of the colony at least one year before they first bred. This intensity of observation was not possible in other years and, as a result, the data in Table 11.5 probably underestimate the age at which young birds first visited the colony. Nevertheless, they show that there was considerable individual variation, with males, on average, tending to arrive at a slightly younger age. Two individuals that were first seen when nine years old had both been marked as chicks in the colony, and it is not known where they had been during the intervening years, except that they were not present in their natal colony. Had they been breeding elsewhere? The female involved bred in the natal colony in the same year as she was first observed, but the male behaved as a typical prospector and delayed breeding until the following year, when he was ten years old. Overall, 58% of males and 72% of females visited the colony in the year before they bred for the first time, but based on the study by Porter, the proportion is probably even higher, particularly since the first visit by some was brief and at sites not easily seen from inside the warehouse.

TABLE 11.5 *The age when first recorded in the natal colony (North Shields) of 207 kittiwakes ringed as nestlings and which subsequently bred there.*

	Age (years) at first return											
	1	2	3	4	5	6	7	8	9	10	Total	Mean
Male	3	43	61	23	11	8	1		1		151	3.20
Female	0	12	25	11	3	3	1		1		56	3.43

TABLE 11.6 *Age at first breeding of 207 kittiwakes ringed as nestlings at North Shields.*

						Age (years) at first return						
	1	*2*	*3*	*4*	*5*	*6*	*7*	*8*	*9*	*10*	*Total*	*Mean age*
Males		2	61	47	28	10	2			1	151	3.97
%		1.3	40.4	31.1	18.5	6.6	1.3			0.7		
Females			7	22	15	8	2	1	1		56	4.70
%			12.5	39.3	26.8	14.3	3.6	1.8	1.8			

There was also considerable individual variation in the age at which kittiwakes breed for the first time (Table 11.6). On average, males breed for the first time when almost a year (the difference between the averages is 0.73 of a year) younger than the females. This difference might be related to the higher annual mortality of adult males and hence the need for more male recruits each year.

The two males who bred when two years old both failed to incubate the eggs laid by the female and these breeding attempts failed.

TABLE 11.7 *Number of years between the age at first return to the colony and the age of first breeding for 151 male and 56 female kittiwakes.*

Males				Age at first breeding						
Age at return	2	3	4	5	6	7	8	9	10	Total
1		3								3
2	**2**	30	8	3						43
3		**28**	23	9	1					61
4			**16**	7						23
5				**9**	1	1				11
6					**8**					8
7						**1**				1
8										
9									1	1
Total	2	61	47	28	10	2			1	151
Females				Age at first breeding						
Age at return	2	3	4	5	6	7	8	9	10	Total
2		4	5	3						12
3		**3**	10	10	2					25
4			**7**		3		1			11
5				**2**	1					3
6					**2**	1				3
7						**1**				1
8										
9								1		1
Total		7	22	15	8	2	1	1		56

There is a relationship between the time of first return and the time of first breeding. Obviously, a bird cannot breed before it has returned to the colony, but the data in Table 11.7 show that at least 14% of males and 32% of females were seen in the colony two years before they eventually bred, and 2.7% of males and 11% of females were seen at least three years before.

I have very little information on the age at first breeding by immigrants, since few of these were ringed and many could not be aged with certainty by plumage.

WHO ARE THE RECRUITS?

For many years, there has been an assumption that colonies of seabirds are virtually self-reproducing units or closed populations which produce their own young to replace the adult mortality. This required that all of the young return to the colony of their birth, a behaviour that is called *philopatry*[1], which literally means a love for the place of the father. However, this concept of a colony is clearly incorrect. For example when new colonies are formed, the establishing birds have to be immigrants reared in other colonies. Further, it is evident that new colonies grow rapidly in the years immediately following their establishment, yet in many seabirds, including the kittiwake, young birds are several years old when they breed for the first time. So the first young reared in the new colony are not able to breed there for at least this length of time, again indicating that the early growth of a new colony has to come from the arrival of immigrants reared elsewhere. Both at North Shields and at a colony on Coquet Island, Northumberland, it took 9 and 7 years respectively before the first philopatric individuals bred in the colonies. By those times, several hundred immigrants had recruited to the new colonies (Coulson & Coulson 2008).

The questions which remain to be answered are for how long this immigration into colonies occurs, and when do the young reared in a new colony first contribute substantially to colony growth?

WHERE TO BREED – HOME OR AWAY?

Early in my studies, I wanted to know more about where young kittiwakes breed, as this is a very important factor in understanding colonial organisation. Did they return to their natal area or did some move elsewhere? In the late 1950s, I put a yellow colour-ring on each of 170 kittiwake chicks in nests

---

[1] The term 'natal philopatry' is tautologous, while 'breeding-site philopatry' is a total misuse of philopatry as it is used to indicate the return of adults to the places where they had bred in a previous year and does not take into account where the individuals were hatched.

on Brownsman, an island in the Farne Islands archipelago in Northumberland, north-east England.

Three and four years later, most of these chicks that had survived were either breeding or had selected nesting sites in colonies, and I searched for them in colonies within 100km of Brownsman, using equal effort in relation to the size of the colonies examined. The kittiwakes nesting on Brownsman and nearby Staple Island are separated by about 200m, and both islands had similar numbers of nests, yet 2.3 times as many of the marked birds were seen on Brownsman than on the neighbouring island (Table 11.8), suggesting that many of the young kittiwakes had made a distinction between the two colonies, despite their closeness. Relatively few of these marked chicks were found on Inner Farne, 2km away, and only one was found at the (then) next nearest colony at Dunstanburgh, some 14km away, while I found none elsewhere, although several other colonies were searched. These observations suggested that many kittiwakes were philopatric and that the dispersal from the colony followed a simple decreasing 'decay' curve, with far fewer occurring as the distance from the natal colony increased.

Later, this same pattern was obtained, this time studying the chicks colour-ringed at North Shields, where many were recorded in the natal colony, with decreasing numbers found in nearby colonies, and only a single individual seen between 50 and 100km from North Shields.

Some years later, I received a few sightings of birds reared at North Shields breeding much further away, and a Farne Island kittiwake was reported breeding on the Scilly Islands, off the south-west coast of England. These records were inconsistent with the 'decay curve' pattern which I had found in north-east England, and I needed to make sense of these confusing records.

With over a century of ringing, there are abundant records of young birds of many species returning to breed at their place of birth, and these probably gave rise to the idea that most young birds are philopatric. For years, researchers ringing nestlings have later trapped them as breeding adults in the same colony. At the same time, they failed to appreciate that they may have been missing many other individuals that moved to breed elsewhere. It is much more difficult to find an individual that has moved than one which has returned to its natal colony. In the case of the kittiwake, for years I identi-

TABLE 11.8 *The sighting of colour-marked kittiwake chicks when 3 or 4 years old. Since observations were made in two years, some records probably involved repeated sightings of the same individuals.*

Colony	Distance from Brownsman (km)	Number colour ringed	Number of sightings	Percentage of all sightings
Brownsman	0	170	35	64%
Staple Island	0.2	0	15	27%
Inner Farne	1	0	4	7%
Dunstanburgh	14	0	1	2%
Elsewhere	15–100	0	0	0%

fied all of the young which returned to North Shields and bred there, but to find those which moved would have involved searching over 200 different kittiwake colonies which were within a 1,000km radius (see later). What is more, this search would have to be done when the legs of the kittiwakes were visible (and so this excluded the incubation period), and at the same intensity as that used in the natal colony. This is an impossible task.

Nevertheless, as time passed, I obtained more records of kittiwakes reared at North Shields breeding elsewhere. I found some myself in nearby colonies, but both excitingly and unexpectedly, I received records of individuals breeding in northern France, in Sweden and on Heligoland, which involved individuals moving hundreds of kilometres from North Shields.

Further confusing information became evident within the North Shields colony. Although all young birds reared at North Shields were ringed for 38 years, throughout the study I found that the great majority of new breeding birds were without rings, and therefore had immigrated. From time to time, new breeding birds at North Shields had been ringed elsewhere; six of these had been ringed on the Farne Islands (about 50km to the north), two at Dunbar in south-east Scotland, and two from Norway. So a clearer picture was building up of the origins of recruits to the breeding group at North Shields, but these depended on other people's ringing efforts, and kittiwakes were ringed in only a few colonies. There was still the discrepancy of the 'decay curve' and few marked birds being found breeding in other colonies in north-east England. A new approach to the problem was needed.

In order to obtain an unbiased sample of where young birds recruited as breeders, I turned to the BTO records of ringing recoveries for kittiwakes ringed in Britain and reported by the general public who would not know the place of ringing and so would not be biased in favour of the colony of ringing. Most of these records were birds which had been found dead. This analysis was made with Gabriel Neve de Mevergnies, and we used only those kittiwakes ringed as nestlings and excluded all of those which had been recovered outside the breeding season, April to July. The sample was reduced further to exclude all those that were still too young to breed when recovered, that is, under four years old. This left 145 recoveries that we could use, and we considered the distance they had moved from their natal colony. This sample showed that there was the expected peak of the birds being found dead in or near their colony of birth, and numbers decreased rapidly beyond 50km. To this extent, the data fitted the decay curve obtained in earlier investigations.

There were few recoveries between 50km and 300km, and this was also in agreement with searches I had made in kittiwake colonies up to 200km from North Shields and which had produced no sightings of North Shields birds beyond 100km. But surprisingly, the new analysis showed a second peak of recoveries between 400km and 1,000km from the natal colony (Coulson & Neve de Mevergnies 1992). The distribution of numbers with distance was not a simple decay curve as I had believed, but was bimodal, with 17.5% of the recoveries over 300km from the natal colony (Figure 11.16). Clearly, many

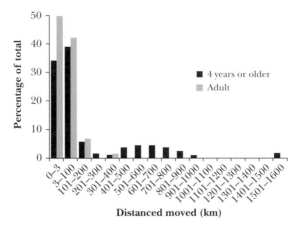

FIG. 11.16   *The distance moved from their natal colony by young kittiwakes when 4 years old or older and reported in the breeding season (black columns). These are compared to birds ringed as adults in a colony and recovered in a subsequent year during the breeding season (grey columns). Note the bimodal pattern of distribution of young kittiwakes when of breeding age and that 17.5 % were recovered more than 300km from their natal colony. Based on 145 and 64 recoveries of individuals ringed as chicks and adults respectively.*

young kittiwakes, perhaps one in every six of those surviving, moved far away from their natal areas to breed and well beyond the distances over which we had searched. Adults ringed while breeding were reported over much shorter distances and these probably did not represent individuals changing colonies, but birds dying while away feeding or their bodies drifting in sea currents before being washed ashore.

A similar bimodal distribution of recoveries has long been known for the Mallard (Thomson 1931) and is called abmigration. Since the kittiwake analysis, similar bimodal patterns have been found for Woodcock (Hoodless & Coulson 1994) and Lapwings reared in Britain (Thompson *et al.* 1994). Figure 11.17 shows the movements between where kittiwakes hatched and where they were reported or presumed to be nesting when at least four years old if this was over 50km from the natal colony. Most of these were found in colonies and only five of these records were in places where no kittiwake colonies existed. In addition, there were two recoveries in Greenland and it is suspected that these may have been recovered in an earlier year by a hunter and no date of finding was given.

There are still some unanswered questions. Why are some individuals intensely philopatric while others move considerable distances to breed? Has this a genetic basis? What role did the sex of the individuals play?

We did not know the sex of these recovered birds and so could not use this method to decide if there were sex differences in the extent of philopatry and emigration. However, we did know the extent to which the two sexes were philopatric at North Shields and these were composed of 82% males (Coulson

FIG. 11.17   *The ringing recoveries of British-ringed kittiwake nestlings when at least four years old and in the breeding season and at least 50km from the natal colony. Data from the British Trust for Ornithology; map reproduced from Coulson & Neve de Mevergnies (1992).*

& Coulson 2008). Very few females returned to their natal colony and 91% (550/603) of female recruits were immigrants. In contrast, 63.5% (430 of 677) of males were immigrants. This greater tendency of males to return and breed near their place of birth is typical of birds and may well be a system evolved to avoid inbreeding (Greenwood 1980). In relation to this last point, at North Shields I never recorded a sister–brother, father–daughter, or son–mother pairing.

A hint about the sex of the individuals which moved long distances comes from the establishment of new kittiwake colonies at considerable distances from the nearest established colony. This occurred in Denmark and again in Spain, when new colonies were established over 500km from the nearest existing one. Both sexes must have been involved in forming these new colonies, so it seems likely that both sexes contribute to the long-distance emigrations of kittiwakes reported from the analysis of ringing recoveries.

It is now evident that a kittiwake colony is by no means a closed population and this conclusion will probably apply to many seabirds when adequate information is available. The low philopatry and high degree of emigration in the kittiwake is of considerable interest for several reasons. First, it indicates considerable flow of genetic material between colonies and regions. A consequence of this is that few subspecies would be expected and that is indeed the case in the kittiwake, with a gradient of size, called a cline, being the main variation detected in either ocean.

Second, the whole mechanism of how young kittiwakes determine the colony in which they eventually breed raises interesting questions. Philopatric individuals do not need to make a choice, but return to the site where they were reared. Of course, how they remember and find the natal colony is of considerable interest, particularly since they probably do not see the colony for several years after fledging. However, with emigration, it becomes evident that many of the recruits are making a choice as to where they will breed. Colour-ringed prospectors often visit several colonies before settling down in one. What attracts them to the one they eventually select? In part, this must be influenced by the size of the colony, as larger colonies need more recruits. It is also likely that in the year or years between visiting colonies and breeding, the presence of many young in the nests might well be an attraction to stay, since their presence would indicate that it is a safe place to breed. This opens up a new area of behavioural research: the identification of what factors make a colony attractive to recruits would add considerably to the management of the species and colonies.

Third, this low level of philopatry has major conservation implications. A colony with frequent low breeding success could be subsidised for many years by immigration from more productive colonies. The threat to the species would be greater if there was widespread, rather than local breeding failure. It is likely that the use of geolocators attached to pre-breeding kittiwakes will supply much relevant information about how young kittiwakes select the colony in which they will eventually breed.

Kittiwake population trends in Britain and Ireland

Comments in books and diaries lead to the suggestion that kittiwake numbers in Britain and Ireland declined during the 19th century, but useful information about numbers does not start to be available until about 1900. At that time, most seabird colonies in Britain were restricted to high cliffs in inaccessible areas, but concerns for their well-being and also for the Grey Seal were being voiced in Parliament. As a result of these initial concerns, a series of bird protection acts were introduced in the first few years of the 20th century. At about the same time, the first Grey Seal Protection Act was passed.

At the start of the 20th century, the numbers of many seabirds reached a low. The Great Skua had been reduced to a few pairs in Scotland and the numbers of colonies of kittiwakes reached an all-time low. My own search of the literature suggested that there were less than 200 pairs of Herring Gulls breeding in England, Wales and southern Scotland in 1900, compared with over 100,000 pairs by 1970. In 1904, the first Fulmars bred on the mainland of Britain (previously they had bred only on the island of St Kilda) (Fisher 1952), and they rapidly and progressively increased in abundance, contributing to the expansion which radically changed the numbers of seabirds during the 20th century. Shooting at colonies during the breeding season became illegal and in Britain, the recovery began for several seabird species as well as the Grey Seal.

Subsequent legislation first reduced, and later effectively banned, the collection of eggs of many species of seabirds. In Britain at the start of the 20th Century, man had suddenly changed from a persecutor of kittiwakes and other seabirds to an active protector. Herring Gulls increased at an incredible rate of about 13% per year from 1900 to 1975. Great Skua numbers recovered remarkably (Furness 1987) as did Grey Seals, both increasing by about 7% per year. The kittiwake increased at about 3% per year for much of the 20th century (Figure 11.18). At first, all of the increases in kittiwake numbers were absorbed by the growth of existing colonies and it was not until the early 1930s, by which time their numbers had increased some threefold, that the first new colonies were formed (Figure 11.19). In 1900, there were only 11 kittiwake colonies known in England, and even by 1950, Fisher & Lockley (1954) could list only 17 colonies there. By the year 2000, I was aware of over 50 separate colonies in England and new ones were still being formed. Similar increases took place in Wales, Scotland, Ireland and the Isle of Man.

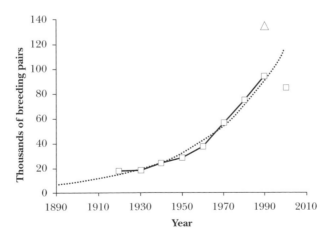

FIG. 11.18 *Estimated numbers of pairs of breeding kittiwakes in England and Wales, 1920 to 2000. Between 1920 and about 1990, the increase closely followed a constant exponential increase of 2.6% per annum, but then numbers declined for the first time by 2000. The open triangle is the total reported in the national census in 1990, but I have been unable to confirm the huge increase reported then for the Flamborough–Bempton cliffs and I believe that the square for that year is the correct estimate of the numbers in 1990. The dotted line represents a 3% annual increase.*

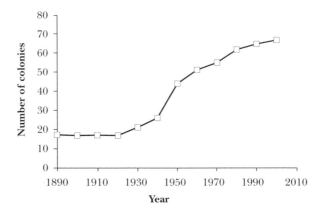

FIG. 11.19   *The number of kittiwake colonies in England and Wales at ten-year intervals from 1890 to 2000. Data for Ireland and Scotland are incomplete for the first half of the 20th century. Recent data from the Joint Nature Conservancy Council's data bank.*

Tracing the recent numbers of kittiwakes in England, the increase is dominated by the large complex of colonies at Flamborough and Bempton in Yorkshire. After reviewing the available information, I have major doubts about some of the numbers of pairs reported for this area (Table 11.9). In 1979, those counting the numbers of kittiwakes at this colony returned figures of 83,000 pairs through the BTO Regional Representative and promised a detailed breakdown of numbers along this stretch of coast, but these were never produced. The numbers returned for this area represented a dramatic increase, far greater than recorded in any other established group of colonies. However, by 2000, the numbers had dropped to 42,659 pairs, indicating an equally dramatic mortality or emigration. After careful consideration and search for more information, I now believe that the 1979 figure supplied was the actual numbers of adult kittiwakes at these colonies, i.e., double the

TABLE 11.9 *The number of breeding pairs of kittiwakes reported from the Flamborough–Bempton cliffs in Yorkshire, 1957–2000. The figures in italics are those after correcting the numbers of pairs.*

Year	Number of pairs	Annual rate of change	Source
1957	22,100		Coulson 1963a
1969	30,800	2.9%	Cramp *et al.* 1974
1979	83,000	9.5%	Coulson 1983
	(41,500)	*(3.0%)*	
1986	83,700	0.1%	Lloyd *et al.* 1991
	(41,850)		
2000	42,659	–4%	Mitchell *et al.* 2004
		(+1%)	
1971–1979		Stable or slight increase	Lloyd *et al.* 1991, accessing data used by Stowe 1982

number of pairs, and there were never anything like 83,000 pairs there at any time. This conclusion has not been made lightly and I supply the evidence for this decision below:

1. The increase between 1969 and 1979 was 169%, which is vast and exceptional for an established colony. No other large colonies have ever increased at a figure of almost 10% per annum over a period of about ten years. Elsewhere, an increase in a large colony of over 3% per year over a similar time period is exceptional.
2. This increase was not accompanied by appreciable increases nearby at the Filey and Scarborough colonies or elsewhere in Great Britain. It was an isolated case.
3. Stowe (1982), when reviewing the status of the kittiwake, recorded that at Flamborough–Bempton the numbers had remained the same or only slightly increased between 1969 and 1979.
4. The figure of 83,700 pairs in 1986 was simply based on the fact that the colonies had remained the same or slightly increased, as suggested by Stowe (1982), and was not a new, total or independent count (as implied by Lloyd *et al.* 1991). Thus it simply repeated the error in the previous count, with a small and arbitrary increase added as suggested by Stowe.
5. The Royal Society for the Protection of Birds (through Keith Clarkson) and I have both been unable to obtain the original counts in 1979 or contact the original counters to check their data.
6. Between 1969 and 2000, it is known that the extent of the colonies changed little, apart from a modest spread at the south of the area. This is surprising if there had been a 169% increase.
7. No decrease occurred in other colonies in north-east England between 1969 and 1979 which could have been the source, through immigration, of the large increase at Flamborough.
8. If a large decrease occurred between 1986 and 2000, it must have resulted from a large increase in the adult mortality rates (which was not reported), and again no decrease in the numbers of pairs of this magnitude was recorded in colonies in north-east England, nor could any colony or groups of colonies be identified which had received some 40,000 additional pairs over this period.
9. It is not without significance that the 1986 count is almost exactly twice that which might have been expected based on earlier and later counts.

It may be that the maximum numbers at this complex of colonies in the 1970s will never be confidently known, but on the evidence available, there must be considerable doubt that there was ever this increase. I suspect that the figure returned for 1979 was the total number of adults and not the number of pairs, thus reducing the numbers to about 41,500 pairs in 1979 and 1986. These numbers are much more consistent with the overall trends in kittiwake numbers in north-east England and would not require the huge increase, followed by a major decrease, neither of which is supported by an independent observer who recorded little change over this period.

From a consideration of these points, I recommend that the record of 83,000 pairs of kittiwakes should be rejected and replaced by 41,500 pairs. When this is done, the decline in kittiwakes in England and Wales from 1985 to 2000 (Figure 11.18) becomes much less, reduced from a loss of 38% to 11% over a 15-year period. The Britain and Ireland decrease is revised from –23% to just over –16% in the same period. Investigations of the numbers of pairs at this large group of colonies are further confused by Mitchell *et al.* (2004) listing more kittiwakes in the Flamborough–Bempton area in 2000 than in the whole Humberside region, of which the former is only a part.

Accepting the revised numbers for the Flamborough area, the pattern of change in the abundance of kittiwakes in England and Wales is appreciably revised and simplified. Kittiwakes have increased at the same rate of 3% per year from 1920 until after 1990, when, for the first time, a decrease occurred (Figure 11.18), but this decrease is just over 6% (0.4% per year) and so much smaller than the 38% decline for England and Wales indicated in Mitchell *et al.* (2004).

However, this is not the only concern about the English section of the 1985–87 census. Lloyd *et al.* (1991) recorded 7,874 pairs of kittiwakes in the Tyne and Wear region and Brown & Grice (2005, p. 386), presumably using the same data source, indicated that 7,700 of these were at Marsden. I know Marsden well, and it has never had more than 5,800 nests at any time, suggesting that the count for the Tyne and Wear region was overestimated by some 2,000 pairs. This is supported by Brown & Grice reporting only 4,074 pairs in the whole of this area and not the 7,874 listed by Lloyd *et al.* (1991). Such a difference would reduce the decline in England during 1985–2000 by a further three percentage points. Yet other discrepancies in Northumberland (colonies at Berwick, Farne Islands, Seahouses, Cullernose Point, Dunstanburgh and Coquet Island) require investigation, as the numbers reported for the county appear to be much too small by several thousand pairs.

CENSUS WORK ELSEWHERE

Elsewhere in Europe, regular census work has only been carried out extensively in recent years and large-scale studies have been initiated in Norway. It appears that most colonies are decreasing in size in northern Norway, but elsewhere in Europe census work is less complete and some colonies have been found to be increasing while others were decreasing in neighbouring areas. Currently the data are too limited to separate long-term changes from short-term fluctuations, but enough information is accumulating to allow better interpretations in future years.

In the Pacific Ocean, there seems to be no co-ordinated and regular census work on Black-legged Kittiwakes, although the Alaska Wildlife Service is accumulating data. The small population of Red-legged Kittiwakes has been decreasing in the Pribilof Islands but has been reported to be increasing on Buldir Island along with similar increases there of Black-legged Kittiwakes. Byrd and Williams (1993) report unpublished data collected by D. Siegel-Causey and based on remains found in archaeological sites, that in the past the Red-legged Kittiwake was both more widespread and more abundant. The low clutch size of this species suggests that it is even more dependent on a high adult survival rate than the Black-legged Kittiwake.

CHAPTER 12

Primary moult in the kittiwake

THERE is considerable information regarding the moult in birds, and much of the earlier information has been reviewed by Palmer (1972). In general, there is considerable variation in the timing of moult between and within species, and many of the differences have not been satisfactorily related to hormone releases (Payne 1972) or to environmental signals.

The kittiwake has one annual total moult of feathers, but a partial moult is said to be responsible for the change from winter plumage into breeding plumage, involving the loss of the dark collar and the spot behind the eye in adult birds. This change can take place within a few days and most individuals have lost their winter plumage by the time they return to the colonies.

The primary feathers are replaced in a descending order over a long period of time, starting during the breeding season and continuing for many months afterwards, and considerable geographical variation occurs in the starting date of the annual primary moult. However, the completion of moult takes place when individuals from different areas are mixed together during their winter distribution, and the breeding areas of birds collected in late autumn and winter and preserved in museums are unknown. As a result, the date by which the primary moult is completed for kittiwakes breeding in different areas cannot be followed. Such information will only be known when enough skins of ringed birds (and thus from known breeding areas) recovered outside the breeding season are accumulated.

The most important part of moult is the replacement of the primary feathers, as this affects flying ability, particularly when the longer outer primary feathers, which determine the length of the wing, are shed. As in many birds, the eleven primaries are moulted in descending order, that is, the moult starts with the inner primary (P1). The small, outermost primary (P11), which is on the bastard wing, is usually ignored in recording the progress of the moult, and the method used to record moult follows that originally proposed by Ginn & Melville (1983). A score of five points is given to the stage of moult for each of the ten main primaries, with a score of zero indicating that moult had not started and a score of 50 that it had been completed and all of the primaries were fully grown. In the kittiwake, many of the primaries are moulted in pairs rather than each at equal time intervals so that, for example, the two inner primary feathers (P1 and P2) are lost at almost the same time, producing an obvious gap between the secondary and primary feathers which is evident in flying birds. Primary moult is usually at the same stage in both wings and in breeding kittiwakes (and several other gulls) in Britain, it commences during incubation (see below). This early onset of moult results in the young being reared while several of the inner five primaries of the parents have been lost and are being progressively replaced by new primary feathers. However, the loss of the long outer primaries, which might be expected to affect flight efficiency, does not occur until long after the young are fledged, and does not coincide with the additional effort required in obtaining food for the young.

Moult scores (for primary moult) were recorded when individual birds were captured for colour-ringing at North Shields and nearby colonies at Marsden and Gateshead, north-east England. The timing of moult in first-year and second-year birds was mainly obtained by netting young birds in a day roost on the south pier at South Shields, where capturing continued during August and September. Captures of adults at the same site gave additional information on the state of moult in breeding birds after colonies were vacated. Since the majority of breeding birds were captured at the nest site where the timing of egg laying was also recorded, it has been possible to relate the state of moult in individuals to their date of laying – something which has not been investigated previously in birds.

Kittiwakes that had been ringed as nestlings and of known age, and had visited the colony in the years before breeding for the first time, were mainly three- or four-year-olds, but included some two-year-old birds that were also captured and examined for moult.

The primary moult of breeding birds progressed at the same rate between moult scores of 5 and 25. The analysis of the data used linear regression between date and moult score and was extrapolated back to a moult score of 1 to give an estimate of when the first primary was lost. This method probably underestimated the actual date of loss of the first primary by one or two days, but this error is small and applied equally to all breeding birds.

MOULT SEQUENCE IN RELATION TO SEX

No difference in the timing or rate of the primary moult was found between the sexes. A sample of 270 breeding males and 241 breeding females showed a difference of less than a half a day in the average start date and the progress of moult was identical, and therefore sex has been ignored in the further interpretation of the progress of the moult.

MOULT SEQUENCE IN RELATION TO AGE

One-year-old birds

On average, one-year-old kittiwakes started their first primary moult on 21 April, which differs by only one day from that estimated by Pennington *et al.* (1994). This start is about 35 days earlier than in breeding adults (Table 12.1). Moult was complete in some one-year-old individuals by late September, suggesting that primary moult lasted for about 130 days. The places of hatching of these one-year-old individuals and those examined by Pennington *et al.* (1994) were not known.

Two-year-old birds

Two-year-old kittiwakes do not breed; many were identified by plumage, and others by their being ringed as nestlings. Primary moult began 18 days later than in first-year birds, but still 17 days earlier than in breeding adults (Table 12.1). Full primary moult in these immature birds was estimated to have been completed in late August, and this was confirmed in a small sample captured in early September when all of the primary feathers were completely grown. This suggests a primary moult period of about 120 days.

Older non-breeding birds

Non-breeding birds in full adult plumage (mainly three or four years old) started their moult about the same time or slightly later by about five days than the breeding birds and about 35 days later than one-year-old birds.

Breeding adults

On average, breeding adults of both sexes started their primary moult on 26 May, while they were incubating eggs.

TABLE 12.1 *The time and rate of primary moult in kittiwakes of different ages and breeding status in north-east England.*

Age	N	Slope (moult score per day)	r	Start of moult	Estimated date of moult completion using rate for score from 1–25
One-year-olds	20	0.47	0.80	21 April	4 Aug
Two-year-olds	71	0.50	0.84	9 May	17 Aug
Three- or four-year olds non-breeders	323	0.42	0.86	31 May	5 Sept
Breeding birds	411	0.42	0.77	26 May	29 Aug

There is an anomalous situation in the moult of kittiwakes over two years old, because the rate of moult up to a score of 25 (half way through moult) would predict that it would be completed in late August or early September (Table 12.1). However, many adult birds examined in Britain during September (including a few known to be breeding at the North Shields colony) were nowhere near completing their primary moult. The report by Hector Galbraith (in Ginn & Melville 1983) of suspended moult in 12 breeding kittiwakes in late July remains the only record of such. In contrast, the kittiwakes examined by Pennington *et al.* (1994), Smith (1988) and in the North Shields study did not find evidence that primary moult had been suspended, and those reported by Galbraith would seem to be an exceptional occurrence. However, it is evident that the second half of the moult of the primary feathers proceeds at a much slower rate than that for the inner primaries. The statement by Pennington *et al.* (1994) that their September samples of adults were unrepresentative because the extent of primary moult was less than they had expected was probably unwarranted.

Primary moult and replacement of the larger, outer feathers in the kittiwake is much slower than that of the first five primaries and it is likely that North Shields adults did not complete moult until the end of November, a delay resulting from the five outer primaries being replaced at about half of the speed of the inner feathers. It is not known with certainty when kittiwakes breeding in Britain complete their primary moult, because this occurs after they have left the colonies and are in their pelagic winter distribution and, by that time, they are mixed with birds from other areas. Examination of skins in several European museums showed that no adult kittiwakes in the collections had completed the primary moult before 30 October and this would apply to British birds. This makes the period of moult of adults spread over at least 160 days and this is certainly much longer than that taken by the one-year-old birds.

Primary moult in relation to laying date

The start of egg laying by individual adult kittiwakes at North Shields ranged from late April until mid-June, with most clutches started between 5 and 30 May. The first breeding bird which showed evidence of the onset of primary moult had lost the inner two primaries in both wings on 16 May and had laid the first egg in the clutch on 2 May, i.e., moult started two weeks into incubation.

An initial impression that moult in breeding kittiwakes was related to the time of egg laying was confirmed by grouping the adults studied into five groups according to their laying dates (Table 12.2). All breeding adults (both sexes) began moulting while incubating. The start of moult was clearly corre-lated with the date of laying, and began later when the laying was late. There was a 23-day difference between the start of moult in the earliest and the latest laying birds. However, the start of moult also became earlier relative to the date of laying in the later laying birds, so that while the earliest breeders started moult 14 days after laying, those birds with eggs laid after 8 June started moult only two days later. Thus between the extreme laying groups, there was a difference of 36 days in egg laying, but the difference in the start of moult changed by 23 days.

The initial rate of progress of moult was slower in earlier breeders and as a result, the moult scores of the late breeders had nearly caught up with those laying earlier by the end of July.

It is evident that two factors affect the timing of moult in breeding kitti-wakes, namely the calendar date and the date of egg laying.

Note that the onset of moult in the three- to four-year-old, non-breeding, prospecting kittiwakes visiting the colony (when others of that age are breeding) was 31 May, similar to the overall average for breeding birds and appreciably later than for the one- and two-year-old, immature birds.

TABLE 12.2 *The timing and rate of primary moult in breeding kittiwakes laying on different dates. Moult score between 5–25 only were used to determine the start date and the rate of moult.*

First egg date	N	Slope (score points per day)	Average start of moult	Average start of moult in days after first egg laid
Before 10 May (mean 6 May)	74	0.40 ± 0.025	20 May	14
11–20 May (mean 16 May)	123	0.42 ± 0.013	27 May	11
21–30 May (mean 24 May)	132	0.42 ± 0.016	2 June	9
31 May–8 June (mean 3 June)	50	0.56 ± 0.026	8 June	5
After 8 June (mean 10 June)	32	0.56 ± 0.029	12 June	2

TIMING OF MOULT ELSEWHERE

It is clear that kittiwakes breeding elsewhere have a different moult pattern from those in Britain. Breeding birds in Greenland did not begin primary moult until late July, but there, as elsewhere in the Arctic, egg laying was also appreciably later.

When I examined skins of kittiwakes collected in France and Denmark during winter (mainly individuals found dead on the shoreline), many adults were still growing the long, outer P10 in late December and January and in these cases P9 (which in the fully developed primaries is shorter than P10) was still longer than P10, giving moult scores of 45–49. These birds (despite not having completed growth of the P10) had relatively long wing lengths compared with those of birds breeding in Britain and presumably the majority came from areas within or near the Arctic Circle, where breeding and the onset of moult both started later. Similarly, kittiwakes wrecked in Shetland in January and February 1993 had still not completed primary feather moult (Weir *et al.* 1996). If these kittiwakes took the same length of time for primary moult as British birds (160 days), then the moult in these kittiwakes may not have started until August, which is near the time of the young fledging rather than during incubation. Do kittiwakes in the far north start moult even later in relation to their breeding cycle?

In the Pacific, the subspecies *R.t. pollicaris* breeds and moults late. From examination of museum material in the Natural History Museum (London), the New York Natural History Museum and the Michigan University collections, no adult kittiwake which had been collected in June in Alaska had started primary moult, and the onset of primary moult was not detected until late July. In contrast, first-year birds in Alaska started moult in late June and moult was well advanced by mid-August, with scores of 15–25. A series of four kittiwakes collected in Alaska, with labels dated 9 September 1911, showed no primary moult and two others with the same date showed the loss of only the two inner primaries. I suspect that the dates on these were incorrect and referred to when they were included in the collection and not the date of shooting.

Johnston (1961) failed to record primary moult in kittiwakes in Alaska while they were rearing young, and so the onset of moult within the Pacific area may be even later in relation to the time of egg laying than that of the Atlantic subspecies near their southern limits. This is supported by information collected by Harrington-Tweit (1979), who found some adult kittiwakes that had died on the Californian coast were still growing the outer primaries in early March, while Howell & Corben (2000) in the same area recorded individual adults that had not completed their primary moult in April. The ages and breeding areas of these birds examined in March and April were unknown, but they were certainly at least in their third year (an age when in Britain the timing of moult was very similar to that of breeding adults). Howell & Corben (2000) also examined museum specimens in the Californian

Academy museum and found that this late completion of moult was evident in specimens that had been collected in other years.

My own measurements of the speed of the first half of moult in kittiwakes in the Pacific area were similar to that in the Atlantic subspecies, taking about two days for each additional unit on the 0–50 scale of the moult score. There is clearly a need for more information on the timing and progress of the primary moult of kittiwakes in the Pacific Ocean, as it seems to differ from that in the Atlantic subspecies or starts very much later in Arctic waters.

Information from both the Atlantic and Pacific Oceans suggests that the completion of the primary moult often takes longer than would have been expected from the time required for the first half of the moult. This later ending of moult could be caused by one (or both) of two possibilities. The same rate of material is put into the primary growth, but the outer feathers are much longer and larger than the inner primaries and therefore require more material and so take longer to complete their growth. The alternative is that the second half of the primary growth can be slowed if and when kittiwakes in the winter oceanic distribution encounter food shortage and storms. However, opposing this latter possibility is the fact that the moult rate had already declined by September, before the birds had moved into their pelagic zones. Again, one-year-old kittiwakes moult earlier and more rapidly, and did not show a later reduced speed of moult. What is evident is that adult kittiwakes spend more than half of each year with incomplete primary feathers. The absence of the inner primaries probably does not greatly affect flight, but the absence of outer primaries, which occurs when the birds have returned to an oceanic life, must have some adverse effect on the efficiency of flying.

There is little information on the primary moult of the Red-legged Kittiwake. One adult I examined had a moult score of 24 on 13 August, suggesting that the timing of the moult was similar, or perhaps slightly earlier than the Black-legged Kittiwakes breeding in the same area.

There is a clear scope for further investigations and information on the primary moult in the kittiwake breeding in other regions, and particularly in those in the more northern regions of the species' distribution.

CHAPTER 13

Kittiwakes and humans

A BRIEF HISTORY OF KITTIWAKES AND HUMANS

THROUGHOUT most of human history, it seems unlikely that there was much contact with either of the two kittiwake species. In more recent centuries, Vikings occasionally included seabirds and eggs in their diet, as did the Inuit people and others, although the specific use of kittiwakes is uncertain. Low-level exploitation of kittiwakes continued well into the 18th century. Where humans and kittiwakes occurred in proximity, there was localized exploitation of eggs, meat and plumage to satisfy the needs of small human populations, for example at the large Flamborough–Bempton cliffs in Yorkshire, England, and at St Kilda (an isolated group of islands to the west of the Hebrides), the Faeroes, Iceland, Norway and Greenland.

Eventually, seafarers became more numerous and travelled further afield. By the 16th century, exploitation of the Great Auk by Europeans was already well established (Birkhead 1993), and presumably they also exploited other seabirds in colonies which they encountered in the North Atlantic. By 1600, seafarers visiting Newfoundland, and the subsequent colonists, utilised seabirds (Pope 2009). It is likely that humans took relatively few kittiwakes at these times, because only small human settlements occurred nearby, and other cliff-nesting seabirds, such as Common and Brünnich's Guillemots, Fulmars and Gannets were more attractive sources of food. Exploitation of the two species of kittiwakes in the Pacific Ocean is more obscure but no doubt occurred in several areas, utilised where needed by local populations.

Accounts from old documents and books refer to 'millions' and 'tens of thousands' of seabirds at particular sites, but in reality no effort was made to enumerate them and exaggeration was rife. We know little about seabird numbers until well into the 20th century. Even publications of the 1940s showed that we still had an imprecise knowledge of the distribution of kittiwakes in Great Britain. Witherby *et al.* (1943), in their classic work *The Handbook of British Birds*, gave little indication of the existence of extensive kittiwake colonies on the east coast of Scotland, while a few years later, Gibson-Hill (1947) erroneously showed that the main abundance of nesting kittiwakes was restricted to western Scotland and the west and south of Ireland, with few on the North Sea coasts of Scotland and England. It was not until the first national census of kittiwakes, made in 1959, that the main areas of abundance of kittiwakes were identified as being along the North Sea coastline of England and Scotland, including Orkney and Shetland (Coulson 1963a).

For most seabirds, including the kittiwake, the industrial revolution of the 18th and 19th centuries changed matters. Travel became more rapid. Firearm production increased, accuracy improved and hunting guns could be reloaded quickly. The dependence upon sail to reach islands diminished as boats powered by engines were developed, providing faster and more reliable travel, and reducing the importance of wind strength and direction. At the same time, the human population was increasing rapidly, but wildlife protection did not exist except at a few isolated places. Exploitation of seabirds spread from small human communities, often restricted to islands, to larger centres of population. Suddenly, there was more interaction between humans and kittiwakes – and much to the detriment of the kittiwakes.

From the 19th century, there are many contemporary accounts of seabirds being killed in vast numbers while breeding at colonies in Europe. Wagers were made as to how many birds could be shot in one day, and one such bet was won by the slaughter of 500 birds by one marksman. In Scotland, early steam ships were hired to take shooting parties to seabird colonies during the breeding season, and the birds were shot as they poured off the cliffs, disturbed by blasts from the steam-powered sirens on the vessels. In addition, goods were more easily and rapidly transferred from place to place during that period. The availability of and markets for seabird eggs increased, and they were sold in large quantities in London and other cities.

In Newfoundland, and probably elsewhere, kittiwakes were captured at sea by fishermen, and the flesh was used on fishing hooks simply because this was the easiest way to obtain bait to attract fish at sea. Fully grown kittiwakes were also taken for food, and these activities explain why relatively large numbers of ringed kittiwakes from Europe were captured and reported near Newfoundland and Greenland during the first half of the 20th century. Even now, capturing kittiwakes in some areas still persists, but on a much reduced scale, judging by the lower numbers of ringing recoveries reported from Newfoundland and Greenland.

In a few isolated localities, seabirds were a major source of food. However, the inhabitants of St Kilda, off the west coast of Scotland, rarely took kitti-wakes. In the Faeroes and Iceland it would appear that kittiwakes were, in most areas, only a secondary food source, with preference given to Puffins and Common Guillemots. In contrast, kittiwakes were always a target for the resident human populations in Greenland, probably because of the very large numbers of immature individuals from colonies throughout the North Atlantic which collected on that coastline in summer and early autumn, while undergoing the annual moult.

Some species of seabirds suffered from human exploitation more than others. The Great Auk quickly became extinct, and some other auks (and their eggs) and young Gannets tended to be favoured because of their large size. By about 1900, many seabird species in Britain are known to have reached their lowest numbers, and this may have applied to other countries on the western coastline of Europe. Human exploitation during the 19th century almost certainly led to a decline in numbers of kittiwakes in the more southern colonies in Europe, but it is less clear what happened at that time in more northern areas where, in general, human populations were smaller. However, the smaller human populations in the colder land areas around the North Atlantic and in the Pacific Oceans presumably resulted in the impact of humans on kittiwakes being less damaging to their numbers.

PROTECTION

At the beginning of the 20th century, legal protection of seabirds was intro-duced in Britain and became progressively more effective. In other European countries, protection for the kittiwake came later, but the development of legal protection in Britain appears to have been a stimulus for increased protection elsewhere in Western Europe, and breeding birds and their eggs are now legally protected over the whole of the North Atlantic. Linked with protection, the census work which started in Britain in the 1950s has traced the recovery of the kittiwake. In other countries, particularly in Norway, Greenland, Iceland and Russia, extensive counts of breeding kittiwakes have been introduced within the last two decades. Trends in abundance in those areas have yet to be measured over an acceptable period of time, but some suggest a decrease while others hint at an increase. Increases and decreases in nearby colonies have been frequently reported, confusing interpretations of trends. In some northern countries in the North Atlantic and in the western Pacific, the presence of colonies has still to be fully recorded and, of course, census work is still incomplete.

Compared with measuring the actual numbers of kittiwakes, it is much easier to trace the spread and colonisation by kittiwakes within countries in Europe. Colonies were established first in Denmark and then later in Germany (Heligoland), Spain and Portugal, while a colony was formed on a small navigation tower off the south coast of Sweden. It is not clear whether kittiwakes ever ceased to breed in France in the 19th century, but numbers have increased there in the 20th century. At the southern end of their range in the western Atlantic, new colonies of kittiwakes have been established in Nova Scotia and New Brunswick, and appreciable increases and spread have occurred in the St Lawrence River area of Canada.

Compared with the Herring Gull, the rate of increase for the kittiwake in Britain was relatively low and averaged about 3% per annum, but over the whole of the 20th century the cumulative effect has been considerable, with the formation of many new colonies (Figure 11.19). It would appear that this explosion in numbers of kittiwake colonies and total numbers in and to the south of the North Sea did not extend to more northern areas, but data are extremely limited. The population trends of the Black-legged Kittiwake during the 20th century in Greenland, Iceland and over much of its distribution in the Pacific are unknown, but it is evident that some new colonies were formed in all of these areas, while other colonies declined, but for unknown reasons.

GENERAL CONSIDERATIONS

Animals cannot increase in abundance forever. The increase of the kittiwake and most of the other seabirds in Britain (and nearby countries) during the 20th century was a recovery from the effects of human persecution, but now numbers have clearly exceeded those in past centuries. It is likely that the recovery was completed before 1960 and since then, the numbers of kittiwakes in Western Europe have probably exceeded those at any time during the past three hundred years. Yet conservation and protection organisations want more and more individuals of all species, and this is an impossible dream. A point will be reached when the environment can no longer support larger numbers and these will, at some stage, level off, fluctuate or even decline because of overpopulation. It is possible that numbers of seabirds in several areas now exceed those which can be sustained. This may have happened for several seabird species in many areas of Western Europe. However, surprise should not be expressed if numbers at some point in time cease to increase, oscillate or even decline for a time. This characteristic of many populations is to be expected. Some years with breeding failure may indicate that this point is being reached and numbers have entered a fluctuating phase, perhaps caused by delayed density-dependent effects.

Some have suggested that the increase of seabirds in Western Europe during the 20th century was primarily due to the expansion of commercial fishing and the accompanying discards. I believe that this explanation is incorrect. It ignores the fact that humans caused an appreciable decline in the first place, and it does not explain the recovery of the Common Eider, which feeds on bivalve molluscs and small crabs. Kittiwakes seem only recently to have started exploiting fish discards and even now follow fishing boats only in some parts of their range. The evidence suggests that the first stimulus for recovery was protection from human exploitation. The extent of the recovery could be limited by a number of possible factors, and man is one of these, but food availability is the ultimate limit.

Obviously, there had to be enough food to support any increases. At first, food was available because seabird numbers had declined considerably and unexploited food sources still existed, but in later years of the increase, additional food sources were exploited. Large gulls started to feed more inland in fields and at landfill sites, while increased fishing activities by humans both directly and indirectly supplied more food. For example, the size of Cod in the North Sea declined in the early half of the 20th century, presumably because the old, large, predatory individuals were progressively caught in greater numbers. This had the initial effect of making Cod more numerous because the biomass of the species was maintained by smaller individuals. A further consequence of this was that there were more Cod of a size that could be exploited by some seabirds.

As human exploitation of fish stocks intensified, the Cod biomass plummeted, and the earlier concerns regarding overfishing became a reality. Kittiwakes do not feed inland or at landfills, but in some places they take discards from fishing boats if not outcompeted by larger species such as Gannets, Fulmars and Herring Gulls. During the breeding season, kittiwakes primarily feed on sandeels (or Capelin in more northern areas) and until the extensive studies at the end of the 20th century, little was known of how stocks of these fish changed.

Sandeels are regularly used as food by many seabird species and also by Grey Seals and several commercial fish species. They are a key group in the food web of the North Sea. Near the end of the 20th century, the introduction of commercial sandeel fisheries resulted in huge quantities literally being vacuumed from the sea by fishing boats; the sandeels were ground into fishmeal, and used as feed for farm animals, fish farms and even as a fertiliser. Many believed that this activity dramatically affected seabird numbers, but others thought the impact was only localised or insignificant. At some point, logic would lead to the understanding that, as the industry increased, it would eventually have a major effect on the animals that use sandeels as their natural food. The evidence that such effects might already exist was correlative and not robust, but it resulted in governmental action to prevent commercial sandeel fisheries in Shetland and in the Firth of Forth (Scotland). The threat in these areas has been alleviated for the time being, but it has not gone away. Overall, sandeel fisheries are still extensive within the North Sea and the adverse impact on other fish and marine organisms has still to be fully evaluated, but since sandeels form a key group of species in the North Sea food web, overexploitation must be prevented.

PACIFIC AREA

The history of the kittiwake in the North Pacific is mainly obscure. Local exploitation of birds and eggs occurred, and probably still occurs, but details are few, particularly in the north-west region.

Information on seabirds in the North Pacific remained sparse until the *Exxon Valdez* oil tanker ran aground on Bligh Reef in Prince William Sound, Alaska in March 1989, leaking 10.8 million gallons of oil into the sea. This event triggered major concerns about the sizes of wildlife populations and the effects oil spills could have in the area. As a result, assessments were made into the possible environmental impact of future oil spills in the North Pacific. As part of these studies, research on the distribution and numbers of seabirds in Alaska was initiated, including both species of kittiwakes. Much information on the two species of kittiwake in the Pacific was gathered, but it is still too recent and possibly not extensive enough to effectively measure population trends with acceptable precision. Both increases and decreases have been recorded. Emerging information suggests that the breeding success of both species of kittiwake fluctuates considerably and is often low in comparison with the Atlantic situation. Man is unlikely to be responsible for this poor breeding success.

CHANGE IN CLIFF HEIGHT
AND USE OF BUILDINGS

In 1900, kittiwake colonies in countries bordering the North Sea were found on high sea cliffs, mainly over 80m high. The relaxation of human exploitation led to kittiwakes forming many new colonies in the following century (Figure 11.19). By 1950, most high sea cliffs with vertical faces had been colonised, and continued expansion included a shift to the use of lower cliffs. This resulted in cliffs as low as ten metres being used. This trend had two additional effects. First, the lower cliffs tended to be nearer human habitation, perhaps at the edge of a harbour or a coastal town, bringing humans and kittiwakes closer together. Second, a new type of 'cliff' became available to kittiwakes, namely structures built by humans. As far as kittiwakes were concerned, some of these had the simple essentials of a natural cliff; they were at the edge of water in harbours and river estuaries, they had vertical faces preventing access by non-avian predators and there were narrow horizontal ledges which could be used as nesting sites.

In 1931, the first record of kittiwakes nesting on a man-made structure occurred at Granton, near Edinburgh, Scotland, where a few pairs nested on the pier for two years. About 50km away, another colony was established in 1934 on a warehouse at Dunbar, which was used until it was demolished in the 1960s. This colony spread onto a natural cliff in the harbour and onto the ruins of a castle wall built on the cliff, with pairs of kittiwakes nesting between the weathered stone blocks of the fortification. Clearly, these kittiwakes showed no preference for natural or artificial structures.

THE SITUATION ON TYNESIDE, NORTH-EAST ENGLAND

In 1949, kittiwakes began nesting on a riverside building near the mouth of the River Tyne at North Shields, north-east England, where much of my study of kittiwakes was to take place. Following the initial colonisation at North Shields, there was a progressive spread of kittiwakes visiting sites up to 24km upriver (Coulson & MacDonald 1962). In 1965, kittiwakes began nesting at Gateshead, 17km from the sea, on window ledges of a waterside sheet metal factory.

This was to be the start of kittiwakes using many buildings in the Newcastle and Gateshead areas, but one after the other, the buildings they had used for only a few years were demolished. Between 1965 and 2000, no fewer than 15 buildings or man-made structures were used as nesting sites, forming distinct kittiwake colonies on the 20km lower stretch of the River Tyne.

Kittiwakes nested on the quayside Customs sheds at Newcastle, and then (in the most inland of all colonies) on the Co-operative Society flour silos 20km from the sea. These buildings were demolished within a few years of the colonies being established, and some birds moved onto the large Rank-Hovis Baltic Flour Mill at Gateshead. Eventually, numbers there increased to over 300 pairs, with kittiwakes breeding on all four sides of the building, but with most on a long ledge overlooking the river. This building continued in use for many years, but was later converted into an art gallery. Plans were made to move the colony by creating sloping surfaces on all horizontal ledges of the art gallery to prevent nesting, and to build a purpose-built nesting structure nearby.

One proposal was to build a curved 'cliff', cantilevered over the river, to allow droppings from nests to fall into the river. This was to be built immediately adjacent to the old Baltic Flour Mill and, when completed, a third of the nesting ledges on the Baltic would be closed off each year, giving the birds time to move while retaining some undisturbed nesting birds in the immediate vicinity during the changeover. Remains of nests from the original colony would be transferred to the new ledges, and tapes of calling kittiwakes would be used if necessary to attract the displaced birds.

Unfortunately, this proposal was considered too expensive and the time involved too long for the developers, and they decided simply to build a tower with a triangular structure containing a series of ledges. Kittiwakes were totally excluded from Baltic Flour Mill ledges by sealing off the nest sites during one winter. It was assumed that the birds would immediately transfer to the new structure in the following spring, but they did not. The new structure remained unoccupied through March, April and part of May. However, in mid-May, a few kittiwakes were attracted to the tower and bred there, and since then increasing numbers have nested each year. However, over 85% of the kittiwakes displaced from the former Baltic Flour Mill failed to nest on the tower and moved elsewhere. After the tower had been in place for six years, the director of the Baltic Centre art gallery decided that droppings from kittiwakes nesting on the tower were also a problem (as I had foreseen and warned), and the tower was moved about 2km downstream. The tower continues to be used by kittiwakes to this day, although it has never achieved the numbers which had previously nested on the Baltic Flour Mill. This is not the end of the story – by 2010, some kittiwakes eventually succeeded in securing nests on the now-sloping ledges of the Baltic Centre and the numbers nesting there are now increasing. At Røst in Norway, kittiwakes also nest successfully on the sloping roof of a building. Will they be allowed to stay on the Baltic Centre?

Where did the majority of the kittiwakes from the Baltic Flour Mill go? I had colour ringed a number of adults nesting there, so I am able to answer this question. A few moved further downriver and joined yet another kittiwake colony on a riverside building at (what then was) the International Paint factory, on the south bank of the river. The majority moved across the river to Newcastle and joined a small number which had previously started to nest there on riverside buildings. The much increased colonies in Newcastle flourished, but their droppings soiled the buildings. When these were refurbished, the kittiwakes were excluded and they moved again, this time only a few metres onto buildings nearer the Tyne Bridge. As numbers increased to several hundred pairs, the local council was concerned about the droppings damaging the buildings, (including the listed and protected Guildhall) and soiling the streets below, and wanted to exclude the kittiwakes. However, a concerted and vociferous protection lobby opposed such action. Netting was put up over the ledges on some buildings and spikes were placed on the ledges of the Guildhall to deter the kittiwakes. Other sites were left undisturbed.

To date, the outcome is that many kittiwakes have managed to build nests upon the spikes on window ledges of the Guildhall! Some displaced kittiwakes moved onto the Tyne Bridge itself, nesting on ledges, on the substantial support columns and also on the girders supporting the overhead road. It is now possible to join the thousands of vehicles driving over the bridge each day and see kittiwakes nesting within a few metres of the road!

Kittiwakes are increasing around the Tyne Bridge, so what of the future? The City of Newcastle-upon-Tyne Council employees scrub and clean the pavements below the colony each day, and a successful compromise situation exists. In 2011, the council decided to remove the kittiwakes, but they have given no thought as to where the birds will move. Further expansion of the colony onto buildings in streets further away from the river, which is already happening, will create further problems. Having moved from the sea and now from the river, how far will nesting kittiwakes move into the old part of

Newcastle? Perhaps another structure, provided solely for the kittiwakes and coupled with a decision not to allow nesting elsewhere in Newcastle, will have to be an eventual compromise solution.

NESTING ON STRUCTURES ELSEWHERE

Nesting on man-made structures has occurred elsewhere, but not to the extent of that on the River Tyne. Nesting began on old warehouses at Hartlepool in 1960, and when these were demolished, they moved to a nearby wooden pier used to carry seawater to an industrial plant. At Seaham Harbour, just south of Sunderland, and at Bridlington in Yorkshire, kittiwakes nest on buildings within the harbours.

In south-east England, Kittiwakes nested on buildings at Lowestoft and before these buildings were demolished, an artificial 'cliff' of breeze-blocks and ledges was constructed on one of the piers to accommodate the displaced kittiwakes. Both there and at Dover, they also nested on the sides of the piers. Nearby, at Sizewell nuclear power station, several hundred kittiwakes nest on the metal supports of inflow and outflow sea-water pipes which extend out into the sea, but the site is due for decommissioning in the near future. Drilling platforms both in the North Sea and in the Irish Sea have been colonised by kittiwakes but again, these are temporary structures.

Elsewhere in Europe, kittiwakes have nested on buildings in Norway, both at Ålesund and at Røst, while at the former town they later spread onto the rock face of a nearby road cut, which involved them moving inland from the harbour. Perhaps the most unusual kittiwake colony developed in Denmark, within a nature reserve on the island of Hirsholmene. Essentially, Denmark is a low-lying country where cliffs are both few and low. Here the kittiwakes nested on a boulder beach, which included stones up to a metre in diameter. The 'cliff ledge' was formed where one large boulder met the next. As this colony expanded, nesting pairs moved onto the neighbouring sand and marram grass dunes, nesting on a horizontal surface and without a 'cliff' of any kind. That colony could only have existed with human protection, but rats later became established on the island and the kittiwakes abandoned the site.

In the Pacific, kittiwakes also nest on artificial structures at two localities. One is on the superstructure of a wrecked ship on Middleton Island, Alaska, where they also nest on the window ledges of a former radar tower, and have been studied by Scott Hatch and his co-workers.

Have 'strains' of kittiwakes been produced which select buildings for breeding, perhaps by having been reared on such sites? The clear answer to this is no. Over the years, I have accumulated many records of kittiwakes reared and ringed on natural cliffs and which, when mature, nested on buildings, while I have even more records of birds reared on buildings eventually breeding on natural cliff sites. Kittiwakes make no distinction between natural cliffs and artificial 'cliffs'. Essentially, what attracts them is a vertical surface with some narrow ledges, preferably (but not necessarily) over water.

CHEMICAL POLLUTION

Since the introduction of the insecticide DDT during World War II, this and other insecticides have entered the food chain and have had major adverse effects on vertebrates at or near the top of the food chains, including birds of prey and seabirds. In addition, compounds including mercury, cadmium and lead, often in an organic formulation, have been suspected of having adverse, but usually non-lethal effects on birds. Currently, a long list of man-made chemicals has found its way into the oceans, with traces found in several seabird species. To date, the levels of these chemicals reported in kittiwakes have been well below the levels where they could be considered to have adverse effects on survival and breeding. Nevertheless, in the USA, Britain and Norway, regular monitoring of residues of potentially harmful chemicals is taking place, and new chemicals are frequently being added to the appreciable list of substances being monitored in seabirds, including kittiwakes.

RED TIDES AND ALGAL BLOOMS

Algal blooms, including those which cause red tides, occur naturally but some are caused by chemically enriched water entering the seas as run-off from farmland and industry and from sewage outfalls. For example, the red tides produced by *Gonyaulax tamarensis* in north-east England originated in pollution-rich waters in the Firth of Forth in south-east Scotland and then spread down the east coast of England. Saxitoxin, a neurotoxin produced by the algae, killed many seabirds, particularly Shags and terns (Coulson *et al.* 1968a, 1968b, 1978) but few kittiwakes were affected.

Another unidentified but highly lethal neurotoxin, associated with the dumping of sewage at sea, resulted in a high mortality of kittiwakes in north-east England, particularly in 1998 (Coulson & Strowger 1999). In this case, much of the mortality occurred at sea during the breeding season, but most of the bodies drifted offshore and sank after a few days. This type of mortality could easily be overlooked, and the extensive mortality was only reported by the crew of a single fishing boat, while bodies were washed ashore in brief periods of onshore winds.

CLIMATE CHANGE AND GLOBAL WARMING

The current view is that human activity is the main cause of increased carbon dioxide in the atmosphere, and that this and other man-made chemicals in the atmosphere are the most likely cause of increasing environmental temperatures recorded since the industrial revolution. If this is so, there is reason for concern for the future of both humans and kittiwakes. However, it is still a matter of debate whether the potential problems arising from global warming have already produced effects on seabirds, and kittiwakes in particular, because such claims are based on correlations and, as such, do not establish a cause and effect relationship (Coulson 2001b).

There is a disturbing and increasing tendency to assume or suggest that changes in any aspect of the biology of animals and plants must be caused by climate change, which ignores the fact that changes have occurred throughout the evolution of life without the influence of man-made climate change. How many of these claims will stand the test of time?

THE FUTURE

Humans and kittiwakes are likely to interact to an even greater extent in years to come. Where kittiwakes are allowed to nest in proximity to humans will be determined by people and governmental legislation. Popular opinion can change rapidly (consider the incidence of bird flu affecting humans in recent years) in regard to birds nesting on buildings and near communities.

Pollutants and debris released into the oceans are continuing to increase. It may be that some of these will affect the survival and breeding performance of kittiwakes in years to come, even if currently these do not appear to be a problem.

At sea, conflict is likely to increase as human demands for fish continue to grow. In some marine areas, the available marine productivity is already fully exploited and overfishing is occurring. Even if fish stocks do not decline further as a result of sound scientific management, there is already competition for these limited resources. If humans harvest more, at some point other species will encounter increasing food shortage.

Hopefully, the short-sighted approach to exploiting sandeels and Capelin will soon be recognised, and the immediate threat arising from harvesting these fish in huge quantities will be avoided. However, sandeels are still being taken in vast and unsustainable levels within the North Sea. No doubt the same general problems will arise elsewhere again and again. Food obtainable from the sea is a limited resource, and the increases in seabirds and seals recorded in Europe cannot continue forever. There are good reasons to believe that population ceilings have already been reached for some seabirds. These ceilings will be reflected in the decline in some seabird species due to overpopulation, but separating natural changes from those where the species is in decline through human activity is and will remain difficult. Quite apart from increased fishing for human consumption, the situation may soon face conservationists that if we want more Puffins, we will have to accept a reduction in, say, kittiwakes or seals which feed on the same resources.

One area which remains problematic concerns the conditions of the oceans where kittiwakes spend the winter half of the year. Undoubtedly, changes are occurring there which we do not fully understand or have measured in quantitative terms. Several concerns have been raised about the oceans, e.g., oceans becoming more acidic, amount of plastic debris, changed currents, and nutrient inputs. Currently, we know very little about the oceanic food chains because such areas are both difficult to study and have not been considered of major economic importance. The opportunity to link increased numbers of offshore oil and gas explorations with marine biological investigations will, hopefully, be developed, putting far more biologists at key sites far from land. Having offshore stations from which regular recordings can be made will greatly expand our knowledge of the habitat used by oceanic seabirds, including the kittiwake. Already, there are hints that the conditions of wintering areas for kittiwakes have deteriorated, but we do not know the details or the causes. Have auks, such as the Puffin and Little Auk, and petrels such as the Fulmar and storm-petrels been similarly affected?

The Red-legged Kittiwake remains a species of conservation concern. Its numbers are small and it breeds in a very restricted area of the North Pacific in a small number of colonies. It obviously has more specialised feeding requirements than the Black-legged Kittiwake and its current low clutch size and productivity makes the species highly dependent on a very high adult survival rate. Its location in the breeding season does not help to facilitate research on this species and there are still many unknown aspects of its biology remaining to be investigated before effective conservation can be introduced. Its pelagic life outside the breeding season remains a mystery. Currently, conservation does not extend far beyond protection of breeding sites of this species.

However, there must be concern over the future of the Red-legged Kittiwake because it is still not known why it has such a restricted distribution, particularly since humans do not seem to have played a major part in determining its small numbers, nor is there evidence that it is suffering from competition from its numerous relative, the Black-legged Kittiwake.

In the near future, the Black-legged Kittiwake is likely to remain the most abundant gull in the world, but changes in its abundance have occurred in the past, particularly in areas with large human populations. It is difficult to predict the cause and nature of these changes at this point in time and increased studies and monitoring are essential, linked to unbiased interpretations.

The history and methods used at the North Shields colony

A BOUT two kilometres from the wide mouth of the River Tyne and at the edge of Smith's Docks at North Shields, north-east England, where ships were repaired in dry docks, a brewery was built in 1856. The building ceased to be a brewery about 1900, and was converted to a store for materials removed from ships while they were being repaired nearby.

The Brewery Store, as it became known, was situated with one end immediately on the edge of the river. It had a ground floor and four more floors above. On all four sides of the upper four floors were evenly spaced windows, each with a sill nearly a metre long and about 100mm deep. The windows were in two halves and hinged at the sides, which allowed them to be opened into the building. This proved to be incredibly valuable when the kittiwakes started to nest on the sills, as the windows could be opened without pushing off the nests. Had the windows been hinged out-over, opening them would have been impossible once nests were built on the sills.

I discovered four pairs of kittiwakes nesting on the river side of the warehouse in 1949. When Edward White and I in the next few years developed our interest in breeding kittiwakes nearby at Marsden, we also wished to be able to study kittiwakes as individuals and our attention turned to the warehouse colony, which by 1952 had increased to 13 nesting pairs. As it happened, one of the directors of Smith's Docks had an interest in natural history and had already seen the colony. An appointment with him produced a friendly exchange of information about kittiwakes and he willingly granted us access to the building as and when we wanted to visit. The foreman at the warehouse proved to be extremely helpful and moved materials stored in front of the windows to allow access. However, access towards some of the windows was blocked by a variety of substantial objects and reaching them involved considerable athletic skills which were regularly used in the name of science. I doubt if the study would have been successful if the current Health and Safety at Work regulations then existed!

As the colony grew, kittiwakes progressively spread from the end of the building overlooking the river to its longer sides and then to the inland end of the building where the birds on the sites could not see the river. Initially, the kittiwakes only nested as a single pair on each sill, with the exception of the longer ledge in front of the hoist door on the top floor, on which three and later four pairs nested. Over time, more and more sills had two pairs of kittiwakes, with the nests touching each other and also the sides of the window frames.

Smith's Docks Repairers closed down in 1980 and the brewery store was bought by a local business man, Mr Jim Pepper, who traded as Jim Marine Ltd, and the bottom floor was converted to a workshop. Fortunately, access was willingly granted, although at one stage, I had to rent the top floor to continue the kittiwake study. In 1990, the warehouse was sold to a group of developers and planning permission was granted to convert it to a series of flats for residences. However, as part of the planning agreement, the developers had to create alternative nesting sites for the kittiwakes and these were produced in 1990 on the sides of the lifeboat station about one kilometre nearer the river mouth. During the 1990–91 winter, netting was hung over all of the windows and the nesting kittiwakes were excluded when they tried to nest in late February 1991. Frustrated birds stood on the edge of neighbouring quays, often in pairs, and made repeated attempts over several weeks to penetrate the netting and land on the sills, but fortunately never getting entangled. In April, several pairs of the kittiwakes attempted to nest on a ledge just below the top of one of the now disused dry docks about 50m away, but were repeatedly disturbed, probably by dogs and foxes, and deserted the site before nest building could occur.

The expectation was that the kittiwakes would move onto the new ledges on the lifeboat station. A few pairs of kittiwakes began to nest there, but they were not ringed and so were not from the Brewery Store! Many of those from the Brewery Store moved further and started to nest in 1991 on the cliffs at Tynemouth, 2km away, where there was already a small colony. At Tynemouth, many formed the same pairs as in the previous year at the warehouse, but to demonstrate caution about generalising, one female from the North Shields colony moved over 400km in 1991 to Lowestoft in south-east England, and nested with an unmarked male.

Many of the marked kittiwakes continued to breed at Tynemouth, but most of those that survived until 1998 died when there was a very high mortality of kittiwakes in the area caused by a neurotoxin, and the study of these marked birds ceased. Studies on kittiwakes were then switched to the disused Baltic Flour Mill at Gateshead and buildings on the Newcastle side of the river, about 17km from the river mouth, but when development of these buildings in the early part of this century also resulted in the kittiwakes in these colonies being excluded, there was no longer ready access to colonies in the area.

Methods used at North Shields

The initial study at the Brewery Warehouse was to mark kittiwakes and then determine the extent to which their age affected their breeding biology. This involved capturing birds and giving each a unique combination of three coloured leg rings. A policy of minimal disturbance consistent with the aims of the study was employed. Adults were captured only once, and the birds were observed both from inside the building and also outside, using a telescope and without disturbance. Visits were made twice a week during the breeding season to identify the individuals present, their mates, the sites they were using and clutch and brood sizes. The birds were disturbed only to mark the adults in the first place, to mark the first egg in each clutch (only in some years) to establish the order of laying of the eggs, and to ring and weigh the chicks. Eggs laid in the colony were measured on one of two separate visits each year.

Capturing kittiwakes

Having obtained permission to visit and study the North Shields colony, the next problem was how to capture individuals for ringing. This had never been done before and the problem was approached with caution and a concern that in so doing, the birds might respond too stressfully to being handled. The first adults to be marked were captured late in the 1953 breeding season, using a noose placed on the nest which, hopefully, caught the bird around the legs when gently tightened.

After several failed attempts, the first bird so captured caused me major concern. When releasing the noose from the leg, I realised with horror that the other leg was missing! Had I pulled the noose to hard? Examining the bird further showed that there was a healed stump, from an old injury, in place of the missing leg. This was the only one-legged kittiwake I have ever encountered and it had to be the first one I caught!

The method of capturing kittiwakes in future years changed to using a metre length of wire taken from a crated box of oranges obtained from a greengrocer. One end was shaped into a hook about 20mm long. The window where the birds were nesting was opened about a centimetre and the wire slowly fed out towards the bird until it could be hooked around a leg. The response of most kittiwakes to the appearance of the wire was one solely of curiosity, and not of fear. I suppose kittiwakes never encounter snakes and so they have no innate wariness towards things that might resemble them! Often the curious bird tried to grasp the wire in its beak, causing it to make a twanging noise as it vibrated against the window frame, but once the hooked end of the wire was beneath its body, the kittiwake took little notice of the wire. Having hooked the wire around the bird's leg, the bird was slowly dragged into the warehouse while quickly opening and then closing the window at almost the same time. To the other kittiwakes on adjacent sills, a kittiwake had suddenly disappeared out of sight and they were not alarmed, continuing to incubate without even rising from the eggs. Capturing breeding kittiwakes while in the later stages of incubation or when the chicks were small became the standard method. When the bird was released from the door of the building, in almost all cases it had already returned to the nest site by the time I had walked back inside the building to that window containing its nest. I never encountered any adverse effects arising from capturing and marking the kittiwakes and the wire, rather than the noose suggested by Benson & Suryan (1999), proved to be the more satisfactory method. Visiting non-breeding birds were captured by the same method and having developed it in 1954, it remained unchanged throughout the study.

Aging birds

First-year kittiwakes rarely visited the colony and because of their very different plumage, they were easily identified. Many of the second-year kitti-wakes were characterised by having the outer black edge of the 9th primary extending 60–100mm beyond the black on the inner side of the shaft (Coulson 1959). Usually, most of the 11th primary, on the bastard wing, was black and there were often dark-tipped feathers in the wing coverts. Most, but not every second-year kittiwake could be aged correctly. Only some third-year kittiwakes could be separated from second-year or older adults, depending on whether they retained dark-tipped feathers in the wing, but usually the pattern of the black tip on the 9th primary was of no assistance as this was often similar to the full adults.

Because young birds have to grow quickly to achieve flight, young individuals achieve adult size soon after fledging. In some mammals, growth continues slowly for some years which allows aging, or they have teeth which show year-rings in the dentine, but no such aids to aging exist for birds. There was no way of aging fully adult kittiwakes unless they had been ringed as nestlings.

Sexing kittiwakes

All data from museum collections of kittiwakes show that male kittiwakes are larger than females in several respects, but there is always considerable overlap. I made a detailed analysis of the biometrics as a potential means of sexing kittiwakes (Coulson 2009). Weight is an unreliable means of sexing kittiwakes, since the difference between an individual having just fed and when it had not fed for some hours can change the weight by over 20%. Wing length is only a modest guide, and by using this, only 74% of individuals can be sexed with confidence. The best biometric measurement to sex kittiwakes is the distance from the tip of the beak to the back of the head (known as head and bill length), but even here there is a small overlap between the sexes (and also with kittiwakes from different geographical areas), though 94% of individuals in the colony were correctly sexed using this measure. In fact, no biometric method correctly sexes all kittiwakes.

Many studies on other bird species have combined several biometric values together to form a discriminant formula to sex individuals. In most cases this is still not totally successful, and this applies to kittiwakes. Some have included weight, but its use is dubious because weight can vary considerably, even within a day. Several of the methods used by others resulted in somewhat exaggerated success rates, as they were applied to their sample of sexed birds, but success was inevitably lower when used on other samples. Coulson *et al.* (1983b) found in several gull species that there was little to be gained by using several measures in combination, rather than the best single value, and that the extra work involved in taking more values was often not worth the effort.

Two other methods are available to sex birds. The use of DNA samples to sex birds has a high success rate, but this method was not available until late in the North Shields study and is relatively expensive to carry out if many individuals are involved. Jodice *et al.* (2000) used this method to sex kittiwakes with 98% accuracy.

The most satisfactory method of sexing kittiwakes proved to be direct observation of pairs copulating or intensive food begging by the females and the subsequent feeding of the female by the male. I did not encounter errors in using these aspects of behaviour, such as those that might arise from 'reverse mounting', which would have involved the female mounting the male, while intensive begging and 'food squeaking' by the female proved totally reliable. For many breeding kittiwakes, particularly those which bred in several years, sex was confirmed on several occasions. Kittiwakes frequently change partners between years and, for example, one male took six different female partners during his breeding years and so all six mates were also sexed. These, too, took different partners during their lives and so their partners were also sexed. Exploring these links for one male, who died and was sexed by dissection, resulted in the sexing of 96 other kittiwakes. In these ways, the sex of most breeding birds was confirmed many times and with complete confidence. The only exceptions to this were a few individuals which bred for one year only and also some of the visiting non-breeding birds, which did not remain in the colony to breed at a later date, and these individuals had their sex allocated using their head and bill lengths.

WEIGHT

As in many studies on birds, I began studies of weight by using a range of Pesola spring balances, but soon became disillusioned with these because they needed regular calibration and did not have the accuracy and reproducibility I considered necessary, particularly when more than one person was involved in obtaining weights. I soon changed to using electric balances which had much improved accuracy, reading consistently to 0.1g, but all weights were recorded to the nearest gram, with the exception of the weight of newly hatched chicks.

Balances were regularly checked against a test weight to confirm their accuracy and reliability throughout the study, something that is rarely mentioned in other studies relating to the weight of birds. In cases where adults or chicks regurgitated food while being weighed, the weight of this was added to their weight obtained.

To be strictly correct, what I was recording was *mass* rather than *weight*, but I hope I can be forgiven for writing about weight in this book, as the outcome of weighing individuals on balances is a more familiar term to most general readers.

RETURN

Returning birds at the start of a new breeding season tended to visit the colony in the morning, and visits to record birds present were made before midday. The birds present on all sites were recorded twice weekly from return to departure in the autumn. The identity of breeding birds each year was recorded many times during the breeding season and overlooking the presence of a breeding bird in any breeding season was not a source of error.

TIME OF BREEDING

The colony was visited at 3–4 day intervals and more frequently on occasions. The marked adult birds present, the stage of nest building and the presence of eggs or young were recorded for each sill on every visit.

In most years, the first egg in a clutch was marked, so the laying order was known for subsequent measuring. The eggs in two-egg clutches tended to be laid at intervals of two (occasionally three) days, and so the date of the first egg in each clutch was estimated with a potential error of one day, which extended to two days if a one-egg clutch was laid.

DETERMINING EGG VOLUME

It seemed likely that differences in egg size could influence the chance of the egg hatching and also the size of the chick at hatching. Obviously, the investigation of egg size could not be destructive as the main aim of the study on marked birds was to investigate their breeding biology. As a result, I needed a rapid method to measure the size of eggs laid by individual female kittiwakes which did not put the eggs at risk.

The weight of an egg progressively declines during incubation and so this is not a reliable measure of size unless all of the eggs are weighed at the same age following laying, or a correction is applied to the weight for the length of time since they were laid. Also, weight alone does not give any information about the shape of the egg.

Rather than use weight, I decided to use the maximum length and breadth of each egg, which could be measured accurately with Vernier calipers, because these measurements are the most often used to describe egg size in the literature. However, I wanted a single value for the size of the egg and so I needed to be able to estimate the volume from the two linear measurements. To achieve this I had to develop a formula which linked length and breadth to the volume of an egg (Coulson 1963c). In most years I measured the maximum breadth and length of each egg laid in the colony to within two decimal places of a centimetre. These two linear measurements of the egg can readily be converted to the volume of an ellipsoid using the well established formula 1.33π breadth^2 × length/6.

However, the kittiwake egg only approximates this shape, as the maximum breadth is not half way along the length axis, and so a small correction factor was needed. This was achieved by measuring the volume, breadth and length of a series of kittiwake eggs in collections housed in the Hancock Museum at Newcastle. The internal volume of each egg was measured by slowly filling each through the hole created when the egg was blown with 70% alcohol measured from a burette, after first ensuring that there was no dried membrane left within the egg which could trap and retain air when filled. The volume of each egg was calculated from the length and breadth measurements and the difference determined as a proportion of the ellipsoid volume. This correction gives an estimate of the volume of a kittiwake egg expressed in cubic centimetres (cc) or millilitres (ml) of:

$0.4866 \times b^2 \times l$, where b and l are the breadth and length of the egg measured in centimetres.

I did work out the relationship between egg volume and the weight of a fresh egg when laid. The relationship is: weight (g) = $1.081 \times$ volume (ml).

Examination of the eggs of other gull species and terns showed they produced similarly shaped eggs, with a ratio of length to breadth between 1.40–1.43, and the same formula can be applied to other larid species. While the correction factor varied marginally with slight variations in the shape index of individual eggs, it was very accurate in most cases, and any small errors in individual eggs tended to cancel out when taking the average of a sample of eggs.

Eggs vary in the proportion of the length to width and this was measured by the shape index calculated as the breadth × 100/length.

COLOUR-RINGING

Kittiwakes were captured for ringing at the nests or when non-breeders visited the window sills. In the first year of study, celluloid colour rings were used and these faded within a year. Yellow faded to cream, white changed to cream, red changed to grey as did blue. Thereafter, rings made from 'Darvic' thermo-plastic were used in a series of non-fading colours (Coulson 1963b).

In nearly all cases, each unringed breeding adult was captured during incubation and colour-ringed. Each bird was given three colour rings plus a British Trust for Ornithology (BTO) metal ring which had a unique serial number, with two rings placed on each leg. Colour rings were blue, green, lime, orange, red, white and yellow, with each identified in records by the initial letter of the colour. The seven colours gave 686 combinations, with the metal ring always the lower of the two, but placed on either leg. Subsequently black was also used at the Baltic Flour Mill and was called niger (N) to separate it from the shorthand of B for blue. Grey (called S for silver) and brown (called C for chocolate) were tried for a short time, but their use was not continued as they sometimes proved difficult to identify through the soiled glass in the windows, with a dirty white ring being confused with lime and, in poor light, the brown sometimes being mistaken for red. The series was extended by using a striped ring of blue and yellow in the 1960s, and later, combinations which had been used 25 years earlier (by which time the birds carrying them had all died) were reused and created no problems. Prospecting birds were marked using the same system.

I would recommend those setting out on a new colour ringing programme not to use both a green and a lime coloured ring in their system. The two colours worked well for those active in the study and familiar with the colours, but caused confusion when colour-ringed birds were reported by the general public or other ornithologists, as they usually did not distinguish between lime and dark green, reporting both as 'green'.

Until 1971, the chicks were given a single coloured ring representing the year of ringing and also a BTO ring. Thereafter, each chick was given an engraved colour ring and a BTO metal ring. The colour of the engraved ring still denoted year, while the engraved alphanumeric code of two symbols made each chick individually identifiable without having to read the small numbers on the BTO ring. This change increased the number of re-sightings. When chicks with alphanumeric rings returned to the colony and bred, they were not captured and re-ringed in compliance with the policy of minimal disturbance. I now regret this decision, because it prevented being able to relate their growth curves as chicks to their adult weights.

Neither the rings nor ringing caused the kittiwakes any problems. The rings did not injure the legs or slip down over the feet, and the birds did not peck at them. When captured and ringed, virtually all of the incubating birds returned within minutes to the nest and resumed incubation. I could not detect any differences in the breeding performances in the year when the breeding birds were ringed while nesting and those in birds which had been ringed with an alphanumeric ring several years earlier (when they were chicks) and were not captured when breeding for the first time.

Ring loss was not a problem, but a few alphanumeric rings were lost, apparently caused by a weakness being produced in the ring by the engraving of a vertical symbol such as 'I' or 'L'.

GROWTH RATES OF CHICKS

Initially, the growth rate of a sample of chicks was obtained by weighing them on each visit to the colony until they were about 30 days old. Once details of the characteristic growth pattern were obtained, it was found that between 75 and 300g, the daily increase in weight was essentially constant. As a result, in later years, the growth rates were determined by weighing chicks four times within the 75–300g limits and using these to calculate the daily growth rate over this period.

AUTOMATIC RECORDINGS

In the 1960s, the technology of automatic recording of information was not well advanced and few, if any, could operate without a mains power supply. Fortunately I was able to establish a power supply on the top floor of the building and I am grateful to Alf Bowman for the practical work required to achieve this. Two types of automatic records were used in some years. One used a 16mm cinematic camera, which could be modified to take an exposure of the nest and whatever was present there every 40 seconds. This method was limited by not producing an image in darkness, but the timing of dawn and dusk was recorded and so gave a good check of the time. A 100 foot spool of cine-film lasted three days. These photographic records were laborious to analyse, but produced interesting results, including the numbers of adults present, their arrival and departure, behaviour and position on the nest site. Much of this material was analysed by Andy Hodges as a major part of his Ph.D. thesis.

A less laborious method used minute amounts of radioactive material added to the leg-ring. The two members of the pair were given different levels of the radioactive material and so different amounts of radiation were received by the Geiger counter placed above the nest. Levels of radiation were recorded on a moving chart and gave a continuous record of which member of the pair was present at the nest, or whether both were present. Details of this method and the results are given in Chapter 5 and in Coulson & Wooller (1984).

The age (breeding experience) of pairs of kittiwakes breeding at the North Shields colony from 1954 to 1990

Male breeding experience this year (years)

		1	2	3	4	5	6	7	8
	1	**519**	93	45	26	21	10	7	2
	2	81	**154**	55	27	22	10	4	3
	3	53	50	**71**	38	21	14	7	5
	4	33	35	30	**48**	29	19	8	6
	5	32	25	19	22	**28**	19	7	9
	6	11	25	18	11	15	**23**	14	4
Female	7	11	5	16	12	10	6	**15**	6
experience	8	12	7	5	11	6	6	4	**8**
(years)	9	3	7	10	6	7	5	4	3
	10	5	4	8	6	5	7	4	6
	11	4	4	9	2	6	5	4	1
	12		3	5	6	1	8	4	2
	13	2	1	2	1	6	2	5	2
	14	3	2		1	1	1	1	2
	15			1		1	2	1	
	16	4	1						1
	17		1	2					
	18			2	1				
	19				1				
Total		773	417	298	219	179	137	89	60
% with a mate of same experience		67	37	24	22	16	17	17	13

Male breeding experience this year (years)

9	10	11	12	13	14	15	16	17	18	19	Total
4	2	5	1	1	2	2	1				741
2	4	1	2			2					367
4	2	4	1	1	1		1				273
5	4	1			1			1			220
3	1	4		1	1				1		172
7	2		3	2							135
4	6	1	1	3		1					97
6	1	3	1	2	2		1	1			76
6	3	2	2		2	1					61
2	5	2	2	2		1	1				60
4	4	4	1	1							49
1	4	3	2	2	1						42
2	1	4	2	2	1						33
5	1		4	1	1						23
3	4	1		3	1	1					18
	1	2			3	1					13
1						2					6
							2				5
											1
59	45	37	22	21	16	11	6	2	1		2392
10	11	11	9	10	6		5				

Annual variation in productivity of kittiwakes

The changes in kittiwake productivity at Foula (data from R. W. Furness), elsewhere in Shetland (JNCC data), Tyne area (data from D. Turner), Farne Islands (National Trust and J. Steel) and at Marsden and North Shields. Values of 0.8 or higher (critical levels) shown in bold type.

Year	Foula	Shetland	Isle of May	Tyne	N. Shields	Marsden	Farnes
1954					**1.33**	**1.21**	
1955					**1.00**	**1.11**	
1956					**1.00**	**1.22**	
1957					**1.09**	**0.99**	
1958					**1.29**		
1959					**1.61**	**1.41**	
1960					**1.40**		
1961					**1.37**		
1962					**1.26**		
1963					**1.40**		
1964					**1.49**		
1965					**1.47**	**1.32**	
1966					**1.09**		
1967					**1.41**		
1968					**1.22**		
1969					**1.10**		
1970					**1.25**		
1971	**1.28**				**1.13**		
1972	**1.35**				**1.13**		
1973	**1.30**				**1.09**		
1974	**1.20**				**1.10**	**0.97**	
1975	**1.20**				**1.09**	**1.08**	
1976	**1.48**				**1.13**	**1.13**	
1977	**1.41**				**1.00**		
1978	**1.38**				**1.07**		
1979	**1.19**				**1.06**		
1980	**1.22**				**1.08**		
1981	**1.05**				**1.10**		
1982	**1.24**				**1.13**		
1983	**1.00**				**0.95**		
1984	**1.00**				**1.23**		
1985	**1.00**		**1.02**		**1.20**		
1986	**0.88**	0.68	**1.43**		**1.42**		
1987	0.30	0.39	**1.20**		**1.30**		**1.13**
1988	0	0.06	0.77		**1.14**		**1.17**
1989	0.31	0.14	**1.13**		**1.34**		**1.37**

Year	Foula	Shetland	Isle of May	Tyne	N. Shields	Marsden	Farnes
1990	0.50	0.12	0.14		1.15		**1.00**
1991	0.75	0.57	0.24	**1.20**		**1.11**	0.60
1992	**1.39**	**0.96**	0.65	**1.13**		**1.16**	**1.08**
1993	**0.92**	0.70	0.06	**1.09**		**1.15**	0.73
1994	**1.07**	0.75	0.23	**0.99**		**1.21**	**0.88**
1995	**0.92**	0.54	0.39	1.50		**1.10**	**1.15**
1996	0.72	0.61	0.56	0.79		**1.09**	**1.28**
1997	**1.07**	0.28	0.40	0.58		**0.98**	0.72
1998	0.40	0.08	0.02	0.20		**1.17**	0.33
1999	**0.99**	0.71	0.22	0.40		(0.46)	**0.82**
2000	**0.87**	0.48	**0.94**	**1.12**		**0.97**	**0.98**
2001	0	0.01	0.62	**0.89**		**0.89**	0.71
2002	0.20	0.22	0.45	**1.15**		**1.17**	0.79
2003	0.05	0.02	0.76	**0.88**		**0.98**	**0.86**
2004	0	0.07	0.29	0.67		**0.91**	0.10
2005	0.26	0.64	**0.85**	**0.88**		**0.87**	0.63
2006	0.21	0.59	0.47	**1.12**		**1.08**	0.57
2007	0.01	0.37	0.24	**0.86**		**1.11**	0.24
2008	0	0.09	0.23	0.57		0.75	0.32
2009	0.30	0.53	0.70	**1.15**		**1.22**	**1.18**
2010	0.03	0.04				**1.19**	

Data relating to the effect of position in the colony and age

THE breeding performance of female kittiwakes in relation to their breeding experience, position in the colony (edge or centre) and whether they are breeding with the same or a new partner from the previous year. First-time breeding females have been regarded as breeding with a new partner. Unweighted averages are given to remove the effect of age on the breeding performance at the centre or edge of the colony.

A. Females breeding with the same mate as last year.

Date of laying

					Breeding experience of female				
	1	*2*	*3*	*4*	*5*	*6–7*	*8–10*	*Over 10*	*Unweighted mean after first laying*
Centre		15 May	16 May	15 May	16 May	16 May	15 May	15 May	15.4 May
N		79	70	51	41	69	91	86	
Edge		18 May	18 May	18 May	18 May	19 May	19 May	21 May	18.7 May
N		85	65	62	37	57	57	47	

Clutch size

					Breeding experience of female				
	1	*2*	*3*	*4*	*5*	*6–7*	*8–10*	*Over 10*	*Unweighted mean after first laying*
Centre		2.03	2.16	2.15	2.20	2.19	2.27	2.25	2.18
N		79	70	53	41	69	91	86	
Edge		2.08	1.98	1.96	2.08	2.16	2.32	2.11	2.10
N		85	60	62	37	58	57	46	

Productivity (young fledged per pair)

					Breeding experience of female				
	1	*2*	*3*	*4*	*5*	*6–7*	*8–10*	*Over 10*	*Unweighted mean after first laying*
Centre		**1.29**	**1.61**	**1.58**	**1.50**	**1.60**	**1.47**	**1.43**	**1.51**
N		79	70	51	41	69	91	86	
Edge		**1.28**	**1.41**	**1.35**	**1.49**	**1.36**	**1.15**	**1.24**	**1.33**
N		85	65	62	37	57	57	47	

Percentage breeding success

					Breeding experience of female				
	1	*2*	*3*	*4*	*5*	*6–7*	*8–10*	*Over 10*	*Unweighted mean after first laying*
Centre		**64**	**75**	**73**	**68**	**73**	**65**	**64**	**69.6%**
N		79	70	53	41	69	91	86	
Edge		**62**	**71**	**69**	**72**	**63**	**65**	**59**	**65.9%**
N		85	65	85	37	57	57	47	

B. Females making a change of mate from the previous year.

Date of laying

					Breeding experience of female				
	1	*2*	*3*	*4*	*5*	*6–7*	*8–10*	*Over 10*	*Unweighted mean after first laying*
Centre	**16 May**	**16 May**	**14 May**	**15 May**	**16 May**	**16 May**	**15 May**	**21 May**	**16.1 May**
N	213	64	63	51	26	42	31	24	
Edge	**23 May**	**18 May**	**17 May**	**19 May**	**16 May**	**19 May**	**20 May**	**16 May**	**17.9 May**
N	247	63	44	38	27	46	31	13	

Clutch size

					Breeding experience of female				
	1	*2*	*3*	*4*	*5*	*6–7*	*8–10*	*Over 10*	*Unweighted mean after first laying*
Centre	**1.78**	**1.97**	**2.00**	**2.16**	**2.19**	**2.07**	**2.00**	**1.96**	**2.05**
N	213	64	63	51	26	42	31	24	
Edge	**1.80**	**2.03**	**2.11**	**2.03**	**2.04**	**2.16**	**2.03**	**2.12**	**2.07**
N	247	63	44	38	27	46	31	13	

Productivity (young fledged per pair)

				Breeding experience of female					
	1	*2*	*3*	*4*	*5*	*6–7*	*8–10*	*Over 10*	*Unweighted mean after first laying*
Centre	**0.85**	**1.24**	**1.27**	**1.36**	**1.38**	**1.41**	**1.32**	**1.05**	**1.29**
N	213	64	63	51	26	42	31	24	
Edge	**0.88**	**1.21**	**1.19**	**1.24**	**1.17**	**1.20**	**1.13**	**1.07**	**1.17**
N	247	63	44	38	27	46	31	13	

Percentage breeding success

				Breeding experience of female					
	1	*2*	*3*	*4*	*5*	*6–7*	*8–10*	*Over 10*	*Unweighted mean after first laying*
Centre	**48**	**63**	**64**	**63**	**63**	**68**	**66**	**54**	**63.0%**
N	213	64	63	51	26	42	31	24	
Edge	**49**	**61**	**56**	**61**	**57**	**55**	**56**	**50**	**56.6%**
N	247	63	44	38	27	46	31	13	

List of scientific names of species mentioned in the text

MAMMALS

Grey Seal *Halichoerus grypus*

BIRDS

Arctic Tern *Sterna paradisaea*
Black-headed Gull *Larus ridibundus*
Black-legged Kittiwake *Rissa tridactyla*
California Gull *Larus californicus*
Common Eider *Somateria mollissima*
Common Guillemot *Uria aalge*
Common Gull *Larus canus*
Common Tern *Sterna hirundo*
Feral Pigeon *Columba livia*
Fulmar *Fulmarus glacialis*
Gannet *Morus bassanus*
Glaucous Gull *Larus hyperboreus*
Great Auk *Pinguinus impennis*
Great Black-backed Gull *Larus marinus*
Great Skua *Stercorarius skua*
Great Tit *Parus major*
Gyr Falcon *Falco rusticolus*
Herring Gull *Larus argentatus*
Kestrel *Falco tinnunculus*
Lapwing *Vanellus vanellus*
Mallard *Anas platyrhynchos*
Osprey *Pandion haliaetus*
Peregrine Falcon *Falco peregrinus*
Puffin *Fratercula arctica*
Raven *Corvus corax*
Red-legged Kitiwake *Rissa brevirostris*
Ring-billed Gull *Larus delawarensis*
Robin (European) *Erithacus rubecula*

Roseate Tern *Sterna dougallii*
Sandwich Tern *Thalasseus sandvicensis*
Shag *Phalacrocorax aristotelis*
Song Sparrow *Melospiza melodia*
Western Gull *Larus occidentalis*
White-tailed Eagle *Haliæetus albicilla*
Woodcock *Scolopax rusticola*

FISH

Arctic Cod *Boreogadus saida*
Capelin *Mallotus villosus*
Cod *Gadus morhua*
Herring *Clupea harengus*
Lesser Sandeel *Ammodytes marinus*
Northern Lampfish *Stenobrachius leucopsarus*
Pollock *Theragra chalcogramma*
Snake Pipefish *Entelurus aequoraeus*
Sprat *Sprattus sprattus*

References

AEBISCHER, N. J. & COULSON, J. C. 1990. Survival of the kittiwake in relation to sex, year, breeding experience and position in the colony. *Journal of Animal Ecology* 59: 1063–1071.

AEBISCHER, N. J., COULSON, J. C. & COLEBROOK, J. M. 1990. Parallel long-term trends across four marine trophic levels and weather. *Nature* 347: 753–755.

AINLEY, D. G., FORD, R. G., BROWN, E. D., SURYAN, R. M. & IRONS, D. B. 2003. Prey resources, competition, and geographic structure of Kittiwake colonies in Prince William Sound. *Ecology* 84: 709–723.

ANKER-NILSSEN, T., BARRETT, R. T. & KRASNOV, J. V. 1997. Long- and short-term responses of seabirds in the Norwegian and Barents seas to changes in stocks of prey fish. Pp. 683–698 In: *Forage Fishes in Marine Ecosytems.* Proceedings of the Lowell Wakefield Fisheries Symposium. (Ed. Anon). University of Alaska Sea Grant College Program, Report No. 97.

ANKER-NILSSEN, T. & AARVAK, T. 2006. Long-term studies of seabirds in the municipality of Røst, Nordland. Norwegian Institute for Nature Research, NINA Report 133.

ANKER-NILSSEN, T., BARRETT, R. T., BUSTNES, J. O., ERIKSTAD, K. E., FAUCHALD, P., LORENTSEN, S.-H., STEEN, H., STRØM, H., SYSTAD, G. H. & TVERAA, T. 2007. SEAPOP studies in the Lofoten and Barents Sea area in 2006. NINA Report 249.

ARMSTRONG, E. A. 1947. *Bird display and behaviour.* Lindsay Drummond. London.

ARTYUKHIN, Y. B. & BURKANOV, V. N. 1999. *Sea birds and mammals of the Russian far east: a field guide.* AST Publishing (in Russian). Moscow.

BAILEY, R. S., FURNESS, R. W., GAULD, J. A. & KUNZLIK, P. A. 1991. Recent changes in the population of the sandeel *Ammodytes marinus* at Shetland and estimates of seabird predation. *ICES Marine Science Journal* 193: 209–216.

BAIRD, P. H. 1990. Influence of abiotic factors and prey distribution on diet and reproductive success in three seabirds species in Alaska. *Ornis Scandinavica* 21: 224–235.

BAIRD, P. H. 1994. Black-legged Kittiwake (*Rissa tridactyla*). In *The Birds of North America*, No. 92. (Poole A. & Gill, F. Eds). The Academy of Natural Sciences. Philadelphia.

BAIRD, P. H. & GOULD, P. 1986. The breeding biology and feeding ecology of marine birds in the Gulf of Alaska. USDC, NOAA OCSEAP. Final Report. 45: 121–505.

BARRETT, R. T. 1978. The breeding biology of the kittiwakes, *Rissa tridactyla* (L.) in Troms, north Norway. Unpublished thesis for the degree of Canidatus Realium, University of Tromsø, Norway.

BARRETT, R. T. 1985. Further changes in the breeding distribution and numbers of cliff-breeding seabirds in Sør-Varanger, North Norway. *Fauna Norevegica.* Series C, *Cinclus* 8: 35–39.

BARRETT, R. T. 2003. The rise and fall of cliff-breeding seabirds in Sor-Varanger, NE Norway, 1970–2002. *Fauna Norvegica* 23: 35–41.

BARRETT, R. T. 2007a. Egg laying, chick growth and food of kittiwakes *Rissa tridactyla* at Hopen, Svalbard. *Polar Research* 15: 107–113.

BARRETT, R. T. 2007b. Food web interactions in the southwestern Barents Sea: Black-legged Kittiwakes *Rissa tridactyla* respond negatively to an increase in herring *Clupea harengus. Marine Ecology Progress Series* 349: 269–276.

BARRETT, R. T., CHAPDELAINE, G., ANKER-NILSSEN, T., MOSBECH, A., MONTEVECCHI, W. A., REID, J. B., & VEIT, R. R. 2006. Seabird numbers and prey consumption in the North Atlantic. *ICES Journal of Marine Science* 63: 1145–1158.

BARRETT, R. T., JOSEFSEN T. D. & POLDER, A. 2004. Early spring wreck of Black-legged Kittiwakes *Rissa tridactyla* in North Norway, April 2003. *Atlantic Seabirds* 6: 33–45.

BARRETT, R. T. & KRASNOV, Y. V. 1996. Recent responses to changes in stocks of prey species by seabirds breeding in the southern Barents Sea. *ICES Journal of Marine Science* 53: 731–722.

BARRETT, R. T., LORENTSEN, S.-H. & ANKER-NILSSEN, T. 2006. The status of breeding seabirds in mainland Norway. *Atlantic Seabirds* 8: 97–126.

BARRETT, R. T. & TERTITSKI, G. M. 2000. Black-legged kittiwake *Rissa tridactyla* pp 100–103 In *The status of marine birds breeding in the Barents Sea region.* (Anker-Nilssen, T., Bakken, T., Strøm, H., Golovkin, A. N., Bianki, V. V. & Tatarinkova, I. P. Eds). Norsk Polarinst. Rapportser. No. 113, Norwegian Polar Institute. Tromsø.

BECH, C., LANGSETH, I. & GABRIELSEN, G. W. 1999. Repeatability of basal metabolism in breeding female kittiwakes *Rissa tridactyla. Proceedings of the Royal Society B* 266: 2161–2167.

BELOPOLSKII, L. O. 1957. *Ecology of colonial seabirds of the Barents Sea.* Israel Program for Scientific Translations, Jerusalem. (Translated from Russian, 1961).

BENSON, J. & SURYAN, R. M. 1999. A leg-noose for capturing adult Kittiwakes at the nest. *Journal of Field Ornithology* 70: 393–399.

BENSON, J., SURYAN, R. M. & PIATT, J. F. 2003. Assessing chick growth from a single visit to a seabird colony. *Marine Ornithology* 31: 181–184.

BIDERMAN, J. O., DRURY, W. H., HICKLEY, S. & FRENCH, J. B. Jr. 1978. Ecological studies in the north Baring Sea. Annual reports. OCSEAP, Boulder, Colorado.

BIRDLIFE INTERNATIONAL 2004. *Birds in Europe: population estimates, trends and conservation status.* Cambridge, UK: BirdLife International. BirdLife Conservation Series no. 12.

BIRDLIFE INTERNATIONAL 2008. Species factsheet: *Rissa tridactyla.* Downloaded from http://www.birdlife.org on 11/6/2008.

BIRKHEAD, T. 1993. *Great Auk Islands.* T. & A. D. Poyser. London.

BIRKHEAD, T. R. & NETTLESHIP, D. N., 1988. Breeding performance of Black-legged Kittiwakes *Rissa tridactyla* at a small expanding colony in Labrador. *Canadian Field-Naturalist* 102: 20–24.

BLANK, N. & GRUBER, N. 2007. *Impacts of ocean acidification on shelled Pteropods in the Southern Ocean.* Swiss Federal Institute of Technology. Zurich.

BOGDANOVA, M.I., DAUNT, F., NEWELL, M, PHILIPS, R.A., HARRIS, M.P. & WANLESS, S. 2011. Seasonal interactions in the black-legged kittiwake, *Rissa tridactyla*: links between breeding performance and winter distribution. *Proceedings of the Royal Society B*, published on line January 2011.

BOULINIER, T. & DANCHIN, E. 1996. Population trend in kittiwake (*Rissa tridactyla*) colonies in relation to ectoparasite infestation. *Ibis* 138: 326–334.

BRADSTREET, M. S. W. 1976. Summer feeding ecology of seabirds in eastern Lancaster Sound. Report by LGL Ltd, Toronto, environmental research associates for Norlands Petroleums Ltd, Calgary, Alberta.

BRAUNE, B. M. 1987. Comparison of total mercury levels in relation to diet and molt for nine species of marine birds. *Archive of Environmental Contamination and Toxicology*. 16: 217–224.

BREKKE, B. & GABRIELSEN, G. W. 1994. Assimilation efficiency of adult kittiwakes and Brunnich's guillemots fed capelin and Arctic cod. *Polar Biology* 14: 279–264.

BROWN, A. & GRICE, P. 2005. *Birds in England*. T. & A. D. Poyser. London.

BRUN, E. 1979. Present status and trends in populations of seabirds in Norway. In *Conservation of Marine Birds of Northern North America*. (Bartonek, J. C. & Nettleship, D. N. Eds). US Department of Fisheries and Wildlife. Series Report 11.

BTO e-mail blog. 2009. 113 kittiwakes dead at Sabinanigo in Huesca province of Spain (Pyrenees) on 1 Feb. IoM bird among dead.

BULL, J., WANLESS, S., ELSTON, D. A., DAUNT, F., LEWIS, S. & HARRIS, M. P. 2004. Local scale variability in the diet of Black-legged Kittiwakes *Rissa tridactyla*. *Ardea* 92: 43–82.

BURTT, E. H. Jr. 1974. Success of two feeding methods of the Black-legged Kittiwake. *Auk* 91: 827–829.

BURTT, E. H. 1975. Cliff-facing interaction between parent and chick kittiwakes *Rissa tridactyla* in Newfoundland. *Ibis* 117: 241–242.

BYRD, G. V. & WILLIAMS, J. C. 1993. Red-legged Kittiwake. *Birds of North America* No.60. (Poole, A. & Gill, F. Eds). The Academy of Natural Sciences. Washington.

CADIOU, B. 1999. Attendance of breeders and prospectors reflects the quality of colonies in the Kittiwake *Rissa tridactyla*. *Ibis* 141: 321–326.

CAM, E., COOCH, E. G. & MONNAT, J.-Y. 2005. Earlier recruitment or earlier death? On the assumption of equal survival in recruitment studies. *Ecological Monographs* 75: 419–434.

CAM, E., HINES, J. E., MONNAT, J. -Y., NICHOLS, J. D. & DANCHIN, É. 1998. Are adult non-breeders prudent parents? The kittiwake model. *Ecology* 79: 2917–2930.

CAM, E., MONNAT, J.-Y. & HINES, J. E. 2003. Long-term fitness consequences of early conditions in the kittiwake. *Journal of Animal Ecology* 72: 411–424.

CAMPHUYSEN, C. J. & DEVREEZE, F. 2005. De Drieteenmeeuw als broedvogel in Nederland. *Limosa* 78: 65–74.

CAMPHUYSEN, C. J. & LEOPOLD, M. F. 2007. Drieteenmeeuw vestigt zich op meerdere platforms in Nederlandse wateren. *Limosa* 80: 153–156.

CHAPDELAINE, G. & BROUSSEAU, P. 1989. Size and trends of Black-legged Kittiwake (*Rissa tridactyla*) populations in the Gulf of St Lawrence (Quebec), 1974–85. *American Birds* 43: 21–24.

CHARDINE, J. W. 1999. Overview of seabird status and conservation in Canada. *Bird Trends* 7: 1–7.

CHARDINE, J. W. 2002. Geographic variation in the wingtip patterns of Black-legged Kittiwakes. *Condor* 104: 687–693.

CHARNOV, E. L. & KREBS, J. R. 1974. On clutch-size and fitness. *Ibis* 116: 217–219.

CONOVER, M. R., MILLER, D. E. & HUNT, G. L. Jr. 1979. Female-female pairs and other unusual reproductive associations in Ring-billed and California gulls. *Auk* 96: 6–9.

CORTEN, A. 1986. On the causes of the recruitment failure of herring in the central and northern North Sea in the years 1972–1978. *Journal Conseil International Exploration du Mer.* 42: 281–294.

COULSON, J. C. 1959. The plumage and leg colour of kittiwakes and comments on the non-breeding population. *British Birds* 52: 189–196.

COULSON, J. C. 1963a. The status of the Kittiwake in the British Isles. *Bird Study* 10: 147–149.

COULSON, J. C. 1963b. Improved coloured-rings. *Bird Study* 10: 109–111.

COULSON, J. C. 1963c. Egg size and shape in the Kittiwake *Rissa tridactyla* and their use in estimating age composition of populations. *Proceedings of the Zoological Society of London* 140: 211–227.

COULSON, J. C. 1966a. The movements of the Atlantic Kittiwake with special reference to age. *Bird Study,* 13: 107–115.

COULSON, J. C. 1966b. The effects of the pair-bond and age on the breeding biology of the Kittiwake Gull *Rissa tridactyla. Journal of Animal Ecology* 35: 269–279.

COULSON, J. C. 1968. Differences in the quality of birds nesting in the centre and on the edges of a colony. *Nature* 217: 478–479.

COULSON, J. C. 1971. Competition for breeding sites causing segregation and reduced young production in colonial animals. In *N.A.T.O. Symposium on Population Dynamics.* (den Boer P. J. & Gradwell, G. R. Eds). Leiden.

COULSON, J. C. 1972. The pair bond and the breeding success in the Kittiwake, *Rissa tridactyla. Proceedings of the XV International Ornithological Congress,* Leiden: 424–433.

COULSON, J. C. 1974. Kittiwake *Rissa tridactyla.* In Cramp, Bourne and Saunders, *The Seabirds of Britain and Ireland.* Collins. London.

COULSON, J. C. 1976. The Kittiwake. In *The atlas of breeding birds in Britain and Ireland.* (Sharrock, J. T. R. Ed.). T. & A. D. Poyser, Berkhamsted.

COULSON, J. C. 1983. The changing status of the Kittiwake *Rissa tridactyla* in the British Isles, 1969–79. *Bird Study,* 30: 9–16.

COULSON, J. C. 1986a. A new hypothesis for the adaptive significance of colonial breeding in the Kittiwake *Rissa tridactyla* and other sea-birds. *Acta XVIII Congressus Internationalis Ornithologicus,* Moscow. pp. 892–899.

COULSON, J. C. 1986b. Density regulation in colonial sea-bird populations. *Acta XVIII Congressus Internationalis Ornithologicus,* Moscow. pp. 783–791.

COULSON, J. C. 1988. Lifetime reproductive success in the Black-legged Kittiwake *Rissa tridactyla. Acta XIX Congressus Internationalis Ornithologicus.* pp.2140–2147. University of Ottawa Press. Ottawa.

COULSON, J. C. 2001a. Colonial breeding. In *Biology of Marine Birds.* (Schreiber, E. A. & Burger, J. Eds). CRC Press, Boca Raton.

COULSON, J. C. 2001b. Does density-dependent mortality occur in wintering Eurasian Oystercatchers and breeding Black-legged Kittiwakes? *Ibis,* 143: 500–502.

COULSON, J. C. 2002. Black-legged Kittiwake (Kittiwake) *Rissa tridactyla*. In *The migration atlas; movement of the birds of Britain and Ireland.*(Wernham, C. *et al*. Eds). T. & A. D. Poyser. London.

COULSON, J. C. 2009. Sexing Black-legged Kittiwakes by measurement. *Ringing & Migration* 24: 233–239.

COULSON, J.C. 2010. Seasonal and annual body mass changes in breeding and prospecting Black-legged Kittiwakes *Rissa tridactyla*: adaption or food shortage. *Waterbirds* 33: 179–187.

COULSON, J. C., ARMSTRONG, I. H., HAWKEY, P. & HUDSON, M. J. 1978. Further mass sea-bird deaths from paralytic shell-fish poisoning. *British Birds* 71: 58–68.

COULSON, J. C. & COULSON, B. A. 2008. Measuring immigration and philopatry in seabirds; recruitment to Black-legged Kittiwake colonies. *Ibis* 150: 288–299.

COULSON, J. C. & DIXON, F. 1979. Colonial breeding in sea-birds. In *Biology and Systematics of Colonial Organisms.* (Larwood .G.& Rosen, B. R. Eds). Academic Press. London.

COULSON, J. C. & FAIRWEATHER, J. A. 2001. Reduced reproductive performance prior to death in the Black-legged Kittiwake: senescence or terminal illness? *Journal of Avian Biology* 32: 146–152.

COULSON, J. C & HOROBIN, J. M. 1972. The annual re-occupation of breeding sites by the Fulmar. *Ibis* 114: 30–42.

COULSON, J. C. & JOHNSON, M. P. 1993. The attendance and absence of adult Kittiwakes *Rissa tridactyla* from the nest site during the chick stage. *Ibis* 135: 372–378.

COULSON, J. C. & MACDONALD, A. 1962. Recent changes in the habits of the kittiwake. *British Birds* 55: 171–177.

COULSON, J. C. & NEVE DE MEVERGNIES, G. 1992. Where do young Kittiwakes *Rissa tridactyla* breed, philopatry or dispersal. *Ardea* 80: 187–197.

COULSON, J. C. & PORTER, J. M. 1985. Reproductive success of the Kittiwake *Rissa tridactyla*: the roles of clutch size, chick growth rates and parental quality. *Ibis* 127: 450–466.

COULSON, J. C. & PORTER, J. M. 1987. Long-term changes in recruitment to the breeding group and the quality of recruits at a Kittiwake *Rissa tridactyla* colony. *Journal of Animal Ecology* 56: 675–689.

COULSON, J. C., POTTS, G. R., DEANS, I. R. & FRAZER, S. M. 1968a. Exceptional mortality of shags and other sea-birds caused by paralytic shellfish poison. *British Birds* 61: 381–404.

COULSON, J. C., POTTS, G. R., DEANS, I. R. & FRAZER, S. M. 1968b. Dinoflagellate crop in the North Sea; mortality of shags and other sea-birds caused by paralytic shellfish poison. *Nature* 220: 21–27.

COULSON, J. C. & STROWGER, J. 1999. The annual mortality rate of Black-legged Kittiwakes in NE England from 1954 to 1998 and a recent exceptionally high mortality. *Waterbirds* 22: 3–13.

COULSON, J. C. & THOMAS, C. S. 1980. A study of the factors influencing the duration of the pair-bond in the kittiwake gull *Rissa tridactyla*. *Proc. XVI International Congress of Ornithology*, Berlin. 1978.

COULSON, J. C. & THOMAS, C. S. 1983. Mate choice in the Kittiwake Gull. pp.361–376. In *Mate Choice*. (Bateson, P. P. Ed.) Cambridge University Press. Cambridge.

COULSON, J. C., THOMAS, C. S., BUTTERFIELD, J. E. L., DUNCAN, N., MONAGHAN, P. & SHEDDEN, C. 1983b. The use of head and bill length to sex live gulls Laridae. *Ibis* 125: 549–557.

COULSON, J. C. & THOMAS, C. S. 1985a. Changes in the biology of the Kittiwake *Rissa tridactyla*: a 31 year study. *Journal of Animal Ecology* 54: 9–26.

COULSON, J. C. & THOMAS, C. S. 1985b. Differences in the breeding performance of individual Kittiwake Gulls *Rissa tridactyla*. In *Behavioural ecology: ecological consequences of adaptive behaviour.* (Sibly R. M. & Smith, R. W., Eds). British Ecological Society. London.

COULSON, J. C. & WHITE, E. 1955. Abrasion and loss of rings among sea-birds. *Bird Study* 2: 1–44.

COULSON, J. C. & WHITE, E. 1956. The study of colonies of the Kittiwake *Rissa tridactyla* (L.) *Ibis* 98: 63–79.

COULSON J. C. & WHITE E. 1958a. Observations on the breeding of the kittiwake. *Bird Study* 5: 74–83.

COULSON, J. C. & WHITE, E. 1958b. The effect of age on the breeding biology of the Kittiwake *Rissa tridactyla*. *Ibis* 100: 40–51.

COULSON, J. C. & WHITE, E. 1959. The post-fledging mortality of the Kittiwake. *Bird Study* 6: 97–102.

COULSON, J. C. & WHITE, E. 1960. The effect of age and density of the breeding birds on the time of breeding of the Kittiwake *Rissa tridactyla*. *Ibis* 102: 71–76.

COULSON, J. C. & WHITE, E. 1961. An analysis of the factors influencing the clutch size of the Kittiwake. *Proceedings of the Zoological Society of London.* 135: 207–217.

COULSON, J. C. & WOOLLER, R. D. 1976. Differential survival rates among breeding Kittiwake gulls *Rissa tridactyla*. *Journal of Animal Ecology* 45: 205–213.

COULSON, J. C. & WOOLLER, R. D. 1984. Incubation under natural conditions in the Kittiwake Gull *Rissa tridactyla*. *Animal Behaviour* 32: 1204–1215.

CRAMP, S., BOURNE, W. R. P. & SAUNDERS, D. 1974. *The Seabirds of Britain and Ireland*. Collins, London.

CRAMP, S. & SIMMONS, K. E. L. 1983. *The Birds of the Western Palearctic*, Vol III. Oxford. University Press. Oxford.

CULLEN, E. 1957. Adaptations in the kittiwake to cliff nesting. *Ibis* 99: 275–302.

DANCHIN, E. 1992. The incidence of the tick parasite *Ixodes uriae* in kittiwake *Rissa tridactyla* colonies in relation to the age of the colony, and a mechanism of infecting new colonies. *Ibis* 134: 134–141.

DANCHIN, E., BOULINIER, T. & MASSOT, M. 1998. Conspecific reproductive success and breeding habitat selection: implications for the study of coloniality. *Ecology* 79: 2415–2428.

DANCHIN, E. & CAM, E. 2002. Can non-breeding be a cost of breeding dispersal? *Behavioral Ecology and Sociobiology* 51: 153–163.

DARLING, F. F. 1938. *Bird flocks and the breeding cycle*. CUP. Cambridge.

DAUNT, F., BENVENUTI, S., HARRIS, M. P., DALLO'ANTONIA, L., ELSTON, D. A., & WANLESS, S. 2002. Marine foraging strategies of the black-legged kittiwake *Rissa tridactyla* at a North Sea colony: evidence for a maximum foraging range. *Marine Ecology Progress Series* 245: 239–247.

DEL HOYO, J., ELLIOTT, A. & SARGATAL, J. 1996. *Handbook of the Birds of the World*, Volume 3 (Hoatzin to Auks). Lynx Edicions. Barcelona.

DRAGOO, B. K. & SUNDSETH, K. 1993. The status of Northern Fulmars, kittiwakes and murres at St. George Island in Alaska in 1992. Unpublished report. AMNWR 93/10. US Fish and Wildlife Service. Homer. Alaska.

DRAGOO, D. E. 1991. Food habits and productivity of kittiwakes and murres at St George Island, Alaska. Unpublished M.Sc. thesis, University of Alaska, Fairbanks.

ELEY, S. M. & NUTTALL, P. A. 1984. Isolation of an English uukuvirus (family Bunyaviridae). *Journal of Hygiene*, Cambridge 93: 313–316.

ELLIOTT, K. H., JACOBS, S. R., RINGROSE, J., GASTON, A. J. & DAVOREN, G. K. 2008. Is mass loss in Brunnich's guillemots *Uria lomvia* an adaptation for improved flight performance or improved dive performance? *Journal of Avian Biology* 39: 619–628.

ERIKSTAD, K. E. & SYSTAD, G. H. 2009. Extensive monitoring of kittiwakes in northern Norway. *SEAPOP short report* 8: 1–6.

FAIRWEATHER, J. A. & COULSON, J. C. 1995a. The influence of forced site change on the dispersal and breeding of the Black-legged Kittiwake *Rissa tridactyla*. *Colonial Waterbirds* 18: 30–40.

FAIRWEATHER, J. A. & COULSON, J. C. 1995b. Mate retention in the Kittiwake *Rissa tridactyla* and the significance of nest site tenacity. *Animal Behaviour* 50: 455–464.

FIRSOVA, L. V. 1978. Breeding biology of the Red-legged Kittiwake *Rissa brevirostris* (Brunch), and the Common Kittiwake *Rissa tridactyla* (Linnaeus) on the Commander Islands. In *Systematics and biology of rare and little-studied birds*. Zoological Institute of the USSR Academy of Sciences. Leningrad. In Russian.

FISHER, J. 1952. *The Fulmar*. Collins. London.

FISHER, J. & LOCKLEY, R. M. 1954. *Seabirds*. Collins. London.

FITCH, M. 1979. Monogamy, polygamy, and female-female pairs in Herring Gulls. *Proc. Colonial Waterbird Group* 3: 44–48.

FITZGERALD, G. R. & COULSON, J. C. 1973. The distribution and feeding ecology of gulls on the tidal reaches of the Rivers Tyne and Wear. *Vasculum* 58: 29–47.

FORD, R. G., AINLEY, D. G., BROWN, E. D., SURYAN, R. M. & IRONS, D. B. 2007. A spatially explicit optimal foraging model of Black-legged Kittiwake behavior based on prey density, travel distances and colony size. *Ecological Modelling* 204: 335–348.

FREDERIKSEN, M., EDWARDS, M., RICHARDSON, A. J., HALLIDAY, S. & WANLESS, S. 2006. From plankton to top predators: bottom-up control of a marine food web across four trophic levels. *Journal of Animal Ecology* 75: 1259–1268.

FREDERIKSEN, M., HARRIS, M. P., DAUNT, F., ROTHERY, P. & WANLESS, S. 2004a. Scale-dependent climate signals drive breeding phenology of three seabird species. *Global Change Biology* 10: 1214–1221.

FREDERIKSEN, M., WANLESS, S. & HARRIS, M. P. 2004b. Estimating true age-dependence in survival when only adults can be observed: an example with Black-legged Kittiwakes. *Animal Biodiversity and Conservation* 27: 541–548.

FREDERIKSEN, M., WANLESS, S., HARRIS, M. P., ROTHERY, P. & WILSON, L. J. 2004c. The role of industrial fisheries and oceanographic change in the decline of North Sea black legged kittiwakes. *Journal of Applied Ecology* 41: 1129–1139.

FREDERIKSEN, M, WRIGHT, P. J., HEUBECK, M., HARRIS, M. P., MAVOR, R. A. &

WANLESS, S. 2005. Regional patterns of Kittiwake *Rissa tridactyla* breeding success are related to variability in sandeel recruitment. *Marine Ecology Progress Series* 300: 201–211.

FROMENTIN, B. & PLANQUE, J.-M. 1996. *Calanus* and environment in the eastern North Atlantic. II. Influence of the North Atlantic Oscillation on *C. finmarchicus* and *C. helgolandicus. Marine Ecology Progress Series,* 134: 111–118.

FURNESS, R. W. 1987. *The Skuas.* T & A. D. Poyser. London.

FURNESS, R. W. 1990. A preliminary assessment of the quantities of Shetland sandeels taken by seabirds, seals, predatory fish and the industrial fishery in 1981–83. *Ibis* 132: 205–217.

FURNESS, R. W. & BIRKHEAD, T. R. 1984. Seabird colony distributions suggest competition for food supplies during the breeding season. *Nature* 311: 655–656.

FURNESS, R. W., ENSOR, K. & HUDSON, A.V. 1992. The use of fishery waste by gull populations around the British Isles. *Ardea* 80: 104–113.

FURNESS, R. W. & TASKER, M. L. 2000. Seabird-fishery interactions: quantifying the sensitivity of seabirds to reductions in sandeel abundance, and identification of key areas for sensitive seabirds in the North Sea. *Marine Ecology Progress Series* 202: 253–264.

FYHN, M., GABRIELSEN, G. W., NORDOY, E. S., MOE, B., LANSETH, I. & BECH, C. 2001. Individual variation in field metabolic rate of kittiwakes (*Rissa tridactyla*) during the chick rearing period. *Physiological and Biochemical Zoology* 74: 343–355.

GABRIELSEN, G. W., MEHLUM, F. & NAGY, K. A. 1987. Daily energy expenditure and energy utilization of free ranging Black-legged kittiwakes (*Rissa tridactyla*). *Condor* 89: 126–132.

GABRIELSEN, G. W., KLAASSEN, M. & MEHLUM, F. 1992. Energetics of Black-legged Kittiwake (*Rissa tridactyla*) chicks. *Ardea* 80: 29–40.

GARÐARSSON, A. 2006a. Viðkoma ritu sumarið 2005. *Bliki* 27: 23–26.

GARÐARSSON, A. 2006b. Nýlegar breytingar á fjölda íslenskra bjargfugla. *Bliki* 27: 13–22.

GARTHE, S. & HÜPPOP, O. 1994. Distribution of ship-following seabirds and their utilization of discards in the North Sea in summer. *Marine Ecology Progress Series* 106: 1–9.

GASPARINI, J., ROULIN, A., GILL, V. A., HATCH, S. A. & BOULINIER, T. 2006. Kittiwakes strategically reduce investment in replacing clutches. *Proceedings of the Royal Society B* 273: 1551–1554.

GASTON, A. J. & PERIN, S. 1993. Loss of mass in breeding Brunnich's Guillemots *Uria lomvia* is triggered by hatching. *Ibis* 135: 472–475.

GEAR, S. 2007. Seabird breeding success on Foula in 2006. Unpublished report.

GIBSON-HILL, C. A. 1947. *British Seabirds.* Witherby. London.

GILL, V. A. 1999. Breeding performance of Black-legged kittiwakes (*Rissa tridactyla*) in relation to food availability: a controlled feeding experiment. Unpublished M.Sc. thesis. University of Alaska, Anchorage.

GILL, V. A. & HATCH, S. A. 2002. Components of productivity in black-legged kittiwakes *Rissa tridactyla*: response to supplemental feeding. *Journal of Avian Biology* 33: 113–126.

GILL, V. A., HATCH, S. A. & LANCTOT, R. B. 2002. Sensitivity of breeding parameters to food supply in Black-legged Kittiwakes *Rissa tridactyla. Ibis* 144: 268–283.

GINN, H. B. & MELVILLE, D. S. 1983. *Moult in birds.* British Trust for Ornithology. Tring.

GOLET, G. H., IRONS, D. B. & ESTES, J. A. 1998. Survival costs of chick rearing in black-legged kittiwakes. *Journal of Animal Ecology* 67: 827–841.

GOLET, G. H. & IRONS, D. B. 1999. Raising young reduces body condition and fat stores in black-legged kittiwakes. *Oecologia* 120: 530–538.

GOLET, G. H., IRONS, D. B. & COSTA, D. P. 2000. Energy costs of chick rearing in Black-legged kittiwakes (*Rissa tridactyla*). *Canadian Journal of Zoology* 78: 982–991.

GOLET, G. H., SCHMUTZ, J. A., IRONS, D. B. & ESTES, J. A. 2004. Determinants of reproductive costs in the long-lived Black-legged Kittiwake: a multiyear experiment. *Ecological Monographs* 74: 353–372.

GOSLER, A. 1993. *The great tit.* Hamlyn. London.

GÖTMARK, F. 1980. Foraging flights of Kittiwakes – some functional aspects. *Vår Fågelvärld* 39: 65–74.

GREENWOOD, P. J. 1980. Mating systems, philopatry and dispersal in birds and mammals. *Animal Behaviour* 28: 1140–1162.

HAMER, K. C., MONAGHAN, P., UTTLEY, J. D., WALTON, P. & BURNS, M. D. 1993. The influence of food supply in the breeding ecology of Kittiwakes *Rissa tridactyla* in Shetland. *Ibis* 135: 255–263.

HARRINGTON-TWEIT, B. 1979. A seabird die-off on the Washington coast in mid winter 1976. *Western Birds* 10: 49–56.

HARRIS, M. P. 2006. Seabirds and pipefish: a request for records. *British Birds*: 99: 148.

HARRIS, M. P., NEWELL, M., DAUNT, F., SPEAKMAN, J. R. & WANLESS, S. 2008. Snake Pipefish *Entelurus aequoreus* are poor food for seabirds. *Ibis* 150: 413–415.

HARRIS, M. P. & WANLESS, S. 1990. Breeding success of British kittiwakes *Rissa tridactyla* in 1986–88: Evidence for changing conditions in the northern North Sea. *Journal of Applied Ecology* 27: 172–187.

HARRIS, M. P., WANLESS, S. & ROTHERY, P. 2000. Adult survival rates of Shag (*Phalacrocorax aristotelis*), Common Guillemot (*Uria aalge*), Razorill (*Alca torda*), Puffin (*Fratercula arctica*) and Kittiwake (*Rissa tridactyla*) on the Isle of May 1986–96. *Atlantic Seabirds* 2: 133–150.

HARTLEY, C. H. & FISHER, J. 1936. The marine foods of birds in an inland fjord region in West Spitzbergen. *Journal of Animal Ecology* 5: 370–389.

HATCH, S. A., BYRD, G. V., IRONS, D. B. & HUNT, G. L. Jr. 1993a. Status and ecology of kittiwakes (*Rissa triadactyla* and *R. brevirostris*) in the North Pacific. In *The status, ecology, and conservation of marine birds of the North Pacific.* (Vermeer, K. *et al.* Eds). Canadian Wildlife Service Special Publication. Ottawa.

HATCH, S. A., ROBERTS, B. D. & FADELY, B. S. 1993b. Adult survival of Black-legged Kittiwakes *Rissa tridactyla* in a Pacific colony. *Ibis* 135: 247–254.

HELFENSTEIN, F., DANCHIN, E. & WAGNER, R. H. 2004a. Assortative mating and sexual size dimorphism in Black-legged Kittiwakes. *Waterbirds* 27: 350–354.

HELFENSTEIN, F., DANCHIN, E. & WAGNER, R. H. 2004b. Is male unpredictability a paternity assurance strategy? *Behaviour* 141: 675–690.

HELFENSTEIN, F., TIRARD, C., DANCHIN, E. & WAGNER, R. H. 2004c. Low frequency of extra-pair paternity and high frequency of adoption in Black-legged Kittiwakes. *Condor* 106: 150–157.

HELFENSTEIN, F., WAGNER, R. H., DANCHIN, E. & ROSSI, J. M. 2003a. Are adult non-breeders prudent parents? The kittiwake model. *Ecology* 79: 2917–2930.

HELFENSTEIN, F., WAGNER, R. H. & DANCHIN, E. 2003b. Sexual conflict over sperm ejection in monogamous pairs of kittiwakes *Rissa tridactyla. Behavioral Ecology and Sociobiology* 54: 370–376.

HELGASON, L. B., BARRETT, R. T., LIE, E., POLDER, A., SKAARE, J. U. & GABRIELSEN, G. W. 2008. Levels and temporal trends (1983–2003) of persistent organic pollutants (POPs) and mercury (Hg) in seabird eggs from northern Norway. *Environmental Pollution* 155: 190–198.

HELGASON, L. B., POLDER, A., FOREID, S., BAEK, K., LIE, E., GABRIELSEN, G. W., BARRETT, R. T., & SKAARE, J. U. 2009. Levels and temporal trends (1983–2003) of polybrominated diphenyl ethers and hexabromocyclododecanes in seabird eggs from north Norway. *Environmental Technology and Chemistry* 28: 1096–1103.

HEUBECK, M. 1989. Breeding success of Shetland's seabirds: Arctic skua, kittiwake, guillemot, razorbill and puffin. In *Seabirds and sandeels.* Proceedings of Shetland Bird Club Seminar, Lerwick, Shetland, 1988 (Heubeck, M. Ed.). Shetland Bird Club.

HEUBECK, M. 2004. Black-legged Kittiwake *Rissa tridactyla.* In *Seabird populations of Britain and Ireland.* (Mitchell, P. I., Newton, S. F., Ratcliffe, N. & Dunn, T. E. Eds) T. & A. D. Poyser, London.

HEUBECK, M. 2006. SOTEAG ornithological monitoring programme 2006: summary report. Unpublished report, University of Aberdeen.

HEUBECK, M., MELLOR, R. M., HARVEY, P. V. & MAINWOOD, A. R. 1999. Estimating the population size and rate of decline of Kittiwakes *Rissa tridactyla* breeding in Shetland, 1981–97. *Bird Study* 46: 48–61.

HIPFNER, J. M., ADAMS, P. A. & BRYANT, R. 2000. Breeding success of Black-legged Kittiwakes, *Rissa tridactyla,* at a colony in Labrador during a period of low capelin, *Mallotus villosus,* availability. *Canadian Field-Naturalist* 114: 413–416.

HOBSON, K. A. 1993. Trophic relationships among Arctic seabirds: from tissue-dependent stable-isotope models. *Marine Ecology Progress Series* 95: 7–18.

HODGES, A. F. 1974. A study of the biology of the kittiwake *(Rissa tridactyla).* Unpublished Ph. D. thesis. University of Durham.

HODGES, A. F. 1975. The orientation of adult kittiwakes *Rissa tridactyla* at the nest site in Northumberland. *Ibis* 117: 235–240.

HOODLESS, A. N. & COULSON, J. C. 1994. Survival Rates and movements of British and Continental Woodcock *Scolopax rusticola* in the British Isles. *Bird Study*: 41: 48–60.

HOWELL, S. N. G. & CORBEN, C. 2000. Retarded wing molt in Black-legged Kittiwakes. *Western Birds* 31: 123–125.

HUNT, G. L. Jr., EPPLEY, Z., BRUGESON, B. & SQUIBB, R. 1981. Reproductive ecology, food and foraging areas of seabirds nesting in the Pribilof Islands, 1975–1979. U.S. Department of Commerce, NOAA; OCSEAP Final Report 12: 1–258.

HUNT, G. L. Jr. & HUNT, M. W. 1977. Female-Female Pairing in Western Gulls (*Larus occidentalis*) in Southern California. *Science* 196: 1466–1467.

HUXLEY, S. 1942. *Evolution; The modern synthesis.* George Allen & Unwin. London.

ICES. 2007. Report of the Working Group on Seabird Ecology (WGSE), 19–23 March

2007. Barcelona, Spain. ICES CM 2007/LRC: 05. ICES. 2008. Report of the Working Group on Seabird Ecology (WGSE), 10–14 March 2008. Lisbon, Portugal. ICES CM 2008/LRC:05.

IRONS, D. B. 1998. Foraging area fidelity of individual seabirds in relation to tidal cycles and flock feeding. *Ecology* 79: 647–655.

IRONS, D. B., NYSEWANDER, D. R. & TRAPP, J. L. 1987. Changes in colony size and reproductive success of Black-legged Kittiwakes in Prince William Sound, Alaska. Unpublished Technical Report (8A/88/01. US Fish and Wildlife Service. Anchorage, Alaska.

JNCC. 2007. *UK Seabirds in 2006.* Results from the UK Seabird Monitoring Programme. From www.jncc.gov.uk/seabirds.

JODICE, P. G. R., LANCTOT, R. B., GILL, V. A., ROBY, D. D. & HATCH, S. A. 2000. Sexing adult Black-legged Kittiwakes by DNA, behaviour and morphology. *Waterbirds* 23: 405–415.

JODICE, P. G. R., ROBY, D. D., HATCH, S. A., GILL, V. A., LANCTOT, R. B. & VISSER, G. H. 2002. Does food availability affect energy expenditure rates of nesting seabirds? A supplemental-feeding experiment with Black-legged Kittiwakes (*Rissa tridactyla*). *Canadian Journal of Zoology* 80: 214–222.

JODICE, P. G. R., ROBY, D. D., TURCO, K. R., SURYAN, R. M., IRONS, D. B., PIATT, J. F., SHULTZ, M. T., ROSENAU, D. G., KETTLE, A. B. & ANTHONY, J. A. 2006. Assessing the nutritional stress hypothesis: the relative influence of diet quantity and quality on seabird productivity. *Marine Ecology Progress Series* 325: 267–279.

JODICE, P. G. R., ROBY, D. D., TURCO, K. R., SURYAN, R. M., IRONS, D. B., PIATT, J. F., SHULTZ, M. T., ROSENAU, D. G. & KETTLE, A. B. 2008. Growth rates of Black-legged Kittiwake *Rissa tridactyla* chicks in relation to delivery rate, size, and energy density of meals. *Marine Ornithology* 36: 107–114.

JOHNSON, S. R. & BAKER, J. S. 1985. Productivity studies. In *Population estimation, productivity of nesting seabirds and food habits at Cape Pierce and the Pribilof Islands, Bering Sea, Alaska.* (Johnson, S. R. Ed.). OCS study MMS 85–0068. Minerals Management Service, Anchorage, Alaska.

JOHNSTON, D. W. 1961.Timing of Annual Molt in the Glaucous Gulls of Northern Alaska. *Condor* 63: 474–478.

KARNOVSKY, N. J., HOBSON, K. A., IVERSON, S. & HUNT, G. L. Jr. 2008. Seasonal changes in diets of seabirds in the North Water Polynysa multiple-indicator approach. *Marine Ecological Progress Series* 357: 291–299.

KEHOE, F. P. & DIAMOND, A. W. 2001. Increases and expansion of the New Brunswick breeding population of Black-legged Kittiwakes, *Rissa tridactyla. Canadian Field-Naturalist* 115: 349–350.

KEMPENAERS, B., LANCTOT, R. B., GILL, V. A., HATCH, S. A. & VALCU, M. 2006. Do females trade copulation for food? An experimental study on kittiwakes (*Rissa tridactyla*). *Behavioral Ecology* 18: 345–353.

KILDAW, S. D. 1999. Competitive displacement? An experimental assessment of nest site preferences of cliff-nesting gulls. *Ecology* 80: 576–586.

KILDAW, S. D., IRON, D. B. & BUCK, C. L. 2008. Habitat quality and metapopulation dynamics of Black-legged kittiwakes *Rissa tridactyla. Marine Ornithology* 36: 35–45.

KITAYSKY, A. S., WINGFIELD, J. C. & PIATT, J. F. 1999. Dynamics of food availability, body condition and physiological stress response in breeding Black-legged Kittiwakes. *Functional Ecology* 13: 577–584.

KITAYSKY, A. S., WINGFIELD, J. C. & PIATT, J. F. 2001. Corticosterone facilitates begging and affects resource allocation in the black-legged kittiwake. *Behavioral Ecology* 12: 619–625.

KOTZERKA, J., GARTHE, S. & HATCH, S. A. 2010. GPS tracking devices reveal foraging strategies of Black-legged Kittiwakes. *Journal für Ornithologie* 151: 459–467.

KRASNOV, Y. V. & BARRETT, R. T. 1995. Large-scale interactions among seabirds, their prey and humans in the southern Barents Sea. Pp. 443–456 In *Ecology of Fjords and Coastal Waters*. (Skjoldal, H. R., Hopkins, C., Erikstad, K. E. & Leinaas, H. P. Eds). Elsevier. Amsterdam.

KRASNOV, Y. V., BARRETT, R. T. & NIKOLAEVA, N. G. 2007. Status of black-legged kittiwakes (*Rissa tridactyla*), common guillemots (*Uria aalge*) and Brünnich's guillemots (*U. lomvia*) in Murman, northwest Russia, and Varanger, north-east Norway. *Polar Research*, 26:113–117.

LACK, D. 1943. *The Life of the Robin*. Witherby. London.

LACK, D. 1954. *The natural regulation of animal numbers*. Clarendon Press. Oxford.

LACK, D. 1968. *Ecological adaptations for breeding birds*. Methuen. London.

LANCE, B. K. & ROBY, D. D. 1998. Diet and postnatal growth in Red-legged and Black-legged Kittiwakes: An inter-species comparison. *Waterbirds* 21: 375–387.

LANCE, B. K. & ROBY, D. D. 2000. Diet and postnatal growth in Red-legged and Black-legged Kittiwakes: an interspecies cross fostering. *Auk* 117: 1016–1028.

LANGHAM, N. P. E. 1972. Chick survival in terns (*Sterna* spp.) with particular reference to the Common Tern. *Journal of Animal Ecology* 41: 385–395.

LEWIS, S., WANLESS, S., WRIGHT, P. J., HARRIS, M. P., BULL, J. & ELSTON, D. A. 2001. Diet and breeding performance of black-legged kittiwakes *Rissa tridactyla* at a North Sea colony. *Marine Ecology Progress Series* 221: 277–284.

LLOYD, C., TASKER, M. L. & PARTRIDGE, K. 1991. *The status of seabirds in Britain and Ireland*. T. & A. D. Poyser. London.

LLOYD, D. S. 1985. Breeding performance of kittiwakes and murres in relation to oceanographical and meteorological conditions across the shelf of the southeastern Bering Sea. Unpublished M.Sc. thesis. University of Alaska. Fairbanks.

LOCK, A. R. 1987. Increases in the breeding population of Black-legged Kittiwakes *Rissa tridactyla* in Nova Scotia. *Canadian Field-Naturalist* 101: 331–334.

LØNNE, O. J. & G. W. GABRIELSEN. 1992. Summer diet of seabirds feeding in sea-ice covered waters near Svalbard. *Polar Biology* 12: 685–692.

LORENTSEN, S.-H. 2005. Det nasjonale overvåkningsprogrammet for sjøfugl. Resultater til og med hekkesesongen 2005. NINA Rapport 97.

MACLEAN, G. L. 1993. *Roberts' Birds of Southern Africa*. 6th edition. CTP Book Printers. Cape Town.

MANISCALCO, J. M., OSTRAND, W. D., SURYAN, R. M. & IRONS, B. D. 2001. Passive interference competition by glaucous-winged gulls on black-legged kittiwakes: A cost of feeding in flocks. *Condor* 103: 616–619.

MASSARO, M., CHARDINE, J. W., JONES, I. L., & ROBERTSON, G. J. 2000. Delayed capelin (*Mallotus villosus*) availability influences predatory behaviour of large gulls on black-legged kittiwakes (*Rissa tridactyla*), causing a reduction in kittiwake breeding success. *Canadian Journal of Zoology* 78: 1588–1596.

MASSARO, M., CHARDINE, J. W. & JONES, I. L. 2001. Relationships between Black-legged Kittiwake nest-site characteristics and susceptibility to predation by large gulls. *Condor* 103: 793–801.

MAUNDER, J. E. & THRELFALL, W. 1972. The breeding biology of the Black-legged Kittiwake in Newfoundland. *Auk* 89: 789–816.

MAVOR, R. A., HEUBECK, M., SCHMITT, S. & PARSONS, M. 2008. Seabird numbers and breeding success in Britain and Ireland, 2006. Joint Nature Conservation Committee. *UK Nature Conservation*, No. 31. Peterborough.

MAVOR, R. A., PARSONS, M., HEUBECK, M. & SCHMITT, S. 2005. Seabird numbers and breeding success in Britain and Ireland, 2004. Joint Nature Conservation Committee. *UK Nature Conservation*, No.29. Peterborough.

MCLAREN, P. L. & RENAUD, W. E. 1982. Seabird concentrations in late summer along the coasts of Devon and Ellesmere Islands N.W.T. *Arctic* 35: 112–117

MITCHELL, I. 2006. Impacts of climate change on Seabirds. In *Marine Climate change impacts*. Annual Report Card 2006. (Buckley, P. J., Dye S. R., & Baxter J. M. Eds). Online summary reports. MCCIP, Lowestoft, UK.

MITCHELL, P. I., NEWTON, S. F., RATCLIFFE, N. & DUNN, T. E. 2004. *Seabird Populations of Britain and Ireland*. T. & A. D. Poyser. London.

MOE, B., LANGSETH, I., FYHN, M., GABRIELSEN, G. W. & BECH, C. 2002. Changes in body condition in breeding kittiwakes *Rissa tridactyla*. *Journal of Avian Biology* 33: 225–234.

MØLLER, A. P. 2006. Sociality, age at first reproduction and senescence: comparative analyses of birds. *Journal of Evolutionary Biology* 19: 682–689.

MONAGHAN, P. 1992. Seabirds and sandeels: the conflict between exploitation and conservation in the northern North Sea. *Biodiversity and Conservation* 1: 98–111.

MONAGHAN, P., NAGER, R. G. & HOUSTON, D. C. 1998. The price of eggs: increased investment in egg production reduces the offspring rearing capacity of parents. *Proceedings of the Royal Society B* 265: 1731–1735.

MONCADA, A. V. 2008. Unpublished Ph.D. thesis. Sibling rivalry in black-legged kittiwakes (*Rissa tridactyla*). University of Glasgow.

MONTEVECCHI, W. A. & TUCK, L. M. 1987. *Newfoundland birds*. Nuttall Ornithological Club. no. 21. Cambridge, Massachusetts.

MOSS, R., WANLESS, S. & HARRIS, M. P. 2002. How small Northern Gannet colonies grow faster than big ones. *Waterbirds* 25: 442–448.

MULARD, H., AUBION, T., WHITE, J. F., HATCH, S. A. & DANCHIN, E. 2009. Experimental evidence of vocal recognition in young and adult black-legged kittiwakes. *Animal Behaviour* 76: 1855–1861.

MURPHY, E. C., SPRINGER, A. M. & ROSENEAU, D. G. 1991. High annual variability in reproductive success of Kittiwakes (*Rissa tridactyla* L.) at a colony in western Alaska. *Journal of Animal Ecology* 60: 515–534.

NETTLESHIP, D. N. 1974. Seabird colonies and distributions around Devon Island and vicinity. *Arctic* 27: 95–103.

NEWMAN, J., CHARDINE, J. W. & PORTER, J. M. 1998. Courtship feeding and reproductive success in Black-legged kittiwakes. *Waterbirds* 21: 73–80.

NEWTON, I. 1989. *Lifetime reproduction in birds.* Academic Press. London.

NICE, M. M. 1937. Studies on the life history of the Song Sparrow. *Transactions of the Linnaean Society of New York.* 4: 1–247.

NYELAND, J. 2004. Apparent trends in the Black-legged Kittiwake in Greenland. *Waterbirds* 27: 342–349.

O'CONNOR, R. J. 1974. Feeding Behaviour of the Kittiwake. *Bird Study* 21: 185–192.

ORO, D. & FURNESS, R. 2002 Influences of food availability and predation on survival of kittiwakes. *Ecology* 83: 2516–2528.

OSTRAND, W. D., DREW, G. S., SURYAN, R. M. & MCDONALD, L. L. 1998. Evaluation of radio-tracking and strip transect methods for determining foraging ranges of Black-legged Kittiwakes. *Condor* 100: 709–718.

PALMER, R. S. 1972. Patterns of molting. In *Avian biology*, Vol. II Farner, D. S. & King, J. R. (Eds). Academic Press, New York.

PALUDAN, K. 1955. Some behaviour patterns of *Rissa tridactyla. Viddensk Medd. Dansk. Naturh. Foren.* 117: 1–21.

PARSONS, J. 1970. Relationship between egg size and post-hatching chick mortality in the Herring Gull (*Larus argentatus*). *Nature* 228: 1221–1222.

PARSONS, J. 1971. breeding biology of the herring gull *Larus argentatus.* Unpublished Ph.D. thesis. University of Durham.

PARSONS, J. 1976. Factors determining the number and size of eggs laid by the herring gull. *Condor* 78: 481–492.

PAYNE, R. B. 1972. Mechanisms and control of molt. In *Avian biology*, Vol. II Farner, D. S. & King, J. R. (Eds). Academic Press, New York.

PEARSON, T. H. 1967. Aspects of the feeding ecology of certain seabirds. Unpublished Ph.D. thesis. University of Durham.

PEARSON, T. H. 1968. The feeding biology of seabird species breeding on the Farne Islands, Northumberland. *Journal of Animal Ecology* 37: 521–552.

PENNINGTON, M. G., BAINBRIDGE, I. P. & FEARON, P. 1994. Biometrics and primary moult of non-breeding Kittiwakes *Rissa tridactyla* in Liverpool Bay, England. *Ringing & Migration* 15: 33–39.

PENNYCUICK, C. J. 1987. Flight of auks (Alcidae) and other northern seabirds compared with southern precellariiformes: ornithodolite observations. *Journal of Experimental Biology* 128: 335–347.

PLANQUE, J.-M., & FROMENTIN, B. 1996. *Calanus* and environment in the eastern North Atlantic. I. Spatial and temporal patterns of *C. finmarchicus* and *C. helgolandicus. Marine Ecology Progress Series* 134: 101–109.

POPE, P. E. 2009. Early migratory fishermen and Newfoundland's seabird colonies. *Journal of the North Atlantic* 2: 57–70.

PORTER, J. M. 1990. Patterns of recruitment to the breeding group in the kittiwake *Rissa tridactyla. Animal Behaviour* 40: 350–360.

RANKIN, M. N. & DUFFEY, E. A. G. 1948. A study of the bird life of the North Atlantic. *British Birds* 41: supplement 1–42.

REGEHR, H. M., RODWAY, M. S. & MONTEVECCHI, W. A. 1998. Antipredator benefits of nest-site selection in Black-legged Kittiwakes. *Canadian Journal of Zoology* 76: 910–915.

ROBERTS, B. D. & HATCH, S. A. 1993. Behavioral Ecology of Black-Legged Kittiwakes during chick rearing in a failing colony. *Condor* 95: 330–342.

ROFF, D. A. 1992. *The evolution of life histories.* Chapman & Hall. New York.

ROSENEAU, D. G., SPRINGER, A. M., MURPHY, E. C. & SPRINGER, M. I. 1985. Population and trophic studies of seabirds in the northern Bering and eastern Chukchi seas, 1981. US Department of Commerce, NOAA OCSEAP, final report. 30: 1–58.

ROTHERY, P., HARRIS, M. P., WANLESS, S. & SHAW, D. N. 2002. Colony size, adult survival rates, productivity and population projections of Black-legged Kittiwakes, *Rissa tridactyla*, on Fair Isle. *Atlantic Seabirds* 4: 17–28.

SALOMONSEN, F. 1967. *Fuglene på Grønland.* Rhodos, Copenhagen. (The seabirds of Greenland).

SANDVIK, H. 2001. Sexing animals using biometry: Intra-pair comparison is often superior to discriminant functions. *Fauna Norvegica* 21: 11–16.

SANDVIK, H., ERIKSTAD, K. E., BARRETT, R. T., & YOCCOZ, N. G., 2005. The effect of climate on adult survival in five species of North Atlantic seabirds. *Journal of Animal Ecology* 74: 817–831.

SCHMIDT-NEILSEN, K. 1960. The salt secreting gland of marine birds. *Circulation* 21: 955–967.

SCHMIDT-NEILSEN, K. 1979. *Animal Physiology: Adaptation and environment.* Second edition. Cambridge University Press. Cambridge.

SCHNEIDER, D. & HUNT, G. L. Jr. 1987. A comparison of seabird diets and foraging distribution around the Pribilof Islands AK. In *Marine birds: their feeding ecology and commercial fisheries relationships.* (Nettleship, D. N., Sanger, G. A. & Springer, P. F. Eds). Special Publication. Canadian Wildlife Service. Ottawa.

SLUYS, R. 1982. Geographical variation of the Kittiwake, *Rissa tridactyla. Le Gerfaut* 72: 221–230.

SMITH, R. D. 1988. Age and sex-related differences in biometrics and moult of Kittiwakes. *Ringing & Migration* 9: 44–48.

SPRINGER, A., ROSENEAU, D., MURPHY, E., & SPRINGER, M. 1984. Environmental controls of marine food webs and food habits of seabirds in the Eastern Chukchi Sea. *Canadian Journal of Fisheries and Aquatic Science* 41: 1202–1215.

SPRINGER, A. M., ROSENEAU, D. G., DENBY, S. L., LLOYD, D., MCROY, C. P. & MURPHY, E. C. 1986. Seabird responses to fluctuating prey availability in the eastern Bering Sea. *Marine Ecology Progress Series* 32: 1–12.

STORER, R. W. 1987. The possible significance of large eyes in the Red-legged Kittiwake. *Condor* 89: 192–194.

STOREY, A. E., ANDERSON, R. E., PORTER, J. M. & MACCHARIES, A. M. 1992. Absence of Parent-Young Recognition in Kittiwakes: a Re-Examination. *Behaviour* 120:302–323.

STOWE, T. J. 1982. Recent population trends in cliff breeding seabirds in Britain and Ireland. *Ibis* 124: 502–510.

STROWGER, J. 1993. Aspects of the breeding biology of the Kittiwake Gull *Rissa tridactyla* at Marsden Bay, Tyne and Wear. Unpublished M. Sc. thesis. University of Durham.

SURYAN, R. M. & IRONS, D. B. 2001. Colony and population dynamics of black-legged Kittiwakes in a heterogeneous environment. *Auk* 118: 636 649.

SURYAN, R. M., IRONS, D. B. & BENSON, J. 2000 Prey switching and variable foraging strategies of black-legged Kittiwakes and the effect on reproductive success. *Condor* 102: 374–384.

SURYAN, R. M., IRONS, D. B., KAUFMAN, M., BENSON, J., JODICE, P. G. R., ROBY, D. D. & BROWN, E. D. 2002. Short-term fluctuations in forage fish availability and the effect on prey selection and brood-rearing in the Black-legged Kittiwake (*Rissa tridactyla*). *Marine Ecology Progress Series* 236: 273–287.

SWANN, R. L., HARRIS, M. P. & AITON, D. G. 2008. The diet of European Shag *Phalacrocorax aristotelis*, Black-legged Kittiwake *Rissa tridactyla* and Common Guillemot *Uria aalgae* on Canna during the chick-rearing period. *Seabird* 21: 44–54.

SWARTZ, L. G. 1966. Sea-cliff birds. In *Environment of Cape Thompson region, Alaska.* (Wilimovsky, N. J. & Wolfe, J. N. Eds). US Atomic Energy Commission, Oak Ridge, Tennessee.

THOMAS, C. S. & COULSON, J. C. 1988. Reproductive success of Kittiwake Gulls. In *Reproductive Success.* (Clutton-Brock, T. H. Ed.). Chicago University Press.

THOMAS, C. S. 1983. The relationships between breeding experience, egg volume and reproductive success of the kittiwake *Rissa tridactyla*. *Ibis* 125: 567–574.

THOMPSON, P. M. & OLLASON, J. C. 2001. Lagged effects of ocean climate change on fulmar population dynamics. *Nature* 413: 417–420.

THOMPSON, P. S., BAINES, D., COULSON, J. C. & LONGRIGG, G. 1994. Age at first breeding, philopatry and breeding site-fidelity in the lapwing *Vanellus vanellus*. *Ibis* 136: 474–484.

THOMSON, A. L. 1931. On 'abmigration' among the ducks: an anomaly shown by the results of bird-marking. In *Proceedings of the 7th International Ornithological Congress* (de Beaufort, L. F. Ed.). Amsterdam.

THOMSON, D. L., FURNESS, R. W. & MONAGHAN, P. 1998. Field metabolic rates of Kittiwakes *Rissa tridactyla* during incubation and chick rearing. *Ardea* 86: 169–175.

THRELFALL, W. 1968. The helminth parasites of three species of gulls in Newfoundland. *Canadian Journal of Zoology* 46: 827–830.

TJERNBERG, M. & SVENSSON, M. (Eds). 2007. *Artfakta – Rödlistade vertebrater I Sverige.* [Swedish Red data Book of Vertebrates]. Ardatabanken, SLU, Uppsala.

VAN NOORDWIJK, A. J., MCCLEERY, R. H. & PERRINS, C. M. 1995. Selection for the timing of great tit breeding in relation to caterpillar growth and temperature. *Journal of Animal Ecology* 64: 451–458.

WAGNER, R. H., HELFENSTEIN, F. & DANCHIN, E. 2004. Female choice of young sperm in a genetically monogamous bird. *Proceedings of the Royal Society B* 271: S134–S137.

WANLESS, S., FREDERIKSEN, M., WALTON, J. & HARRIS, M. P. 2009 Long-term changes in breeding phenology at two seabird colonies in the western North Sea. *Ibis* 151: 274–285.

WANLESS, S. & HARRIS, M. P. 1992. Activity budgets, diet and breeding success of kittiwakes *Rissa tridactyla* on the Isle of May. *Bird Study* 39: 145–154.

WANLESS, S., HARRIS, M. P., REDMAN, P. & SPEAKMAN, J. R. 2005. Low energy values of fish as a probable cause of a major seabird breeding failure in the North Sea. *Marine Ecology Progress Series* 294: 1–8.

WANLESS, S., MONAGHAN, P., UTTLEY, J. D., WALTON, P. & MORRIS, J. A. 1992. A radio-tracking study of kittiwakes (*Rissa tridactyla*) foraging under suboptimal conditions. In *Wildlife telemetry: remote monitoring of and tracking of animals.* (Priede, I. G. & Swift, S. M., Eds). Ellis Horwood. New York.

WANLESS, S., WRIGHT, P. J., HARRIS, M. P. & ELSTON, D. A. 2004. Evidence for decrease in size of lesser sandeels *Ammodytes marinus* in a North Sea aggregation over a 30-yr period. *Marine Ecology Progress Series* 279: 237–246.

WEIR, D.N., KITCHENER, A.C. & MCGOWAN, R. Y. M. 1996. Biometrics of Kittiwakes *Rissa tridactyla* wrecked in Shetland in 1993. *Seabird* 18: 5–9.

WETLANDS INTERNATIONAL 2006. *Waterbird Population Estimates* – Fourth Edition. Wageningen, The Netherlands.

WHITE, J., LECLAIRE, S., KRILOFF, M., MULARD, H, HATCH, S. A. & DANCHIN, E. 2010. Sustained increase in food supplies reduces brood mate aggression in black-legged kittiwakes. *Animal Behaviour* 79: 1095–1100.

WINTREBERT, C. M. A., ZWINDERMAN, A. H., CAM, E., PRADEL, R. & VAN HOUWELINGEN, J. C. 2005. Joint modelling of breeding and survival in the kittiwake using frailty models. *Ecological Modelling* 181: 203–213.

WITHERBY, H. F., JOURDAIN, F. C. R.,TICEHURST, N. F. & TUCKER, B. W. 1943[C1]. *The handbook of British birds.* Vol 5. Witherby. London.

WOOLLER, R. D. 1973. Studies on the breeding biology of the kittiwake. Unpublished Ph.D. thesis. University of Durham.

WOOLLER, R. D. 1978. Individual vocal recognition in the Kittiwake gull, *Rissa tridactyla* (L.) *Zeitschrift für Tierpsychologie* 48: 68–86.

WOOLLER, R. D. 1980. Repeat laying by kittiwakes *Rissa tridactyla. Ibis* 122: 226–229.

WOOLLER, R. D. & COULSON, J. C. 1977. Factors affecting the age of first breeding of the Kittiwake *Rissa tridactyla. Ibis* 119: 339–349.

WYNNE-EDWARDS, V. C. 1935. On the habits and distribution of birds on the North Atlantic. *Proceedings of the Boston Society of Natural History* 40: 233–340.

WYNNE-EDWARDS, V. C. 1962. *Animal dispersion in relation to social behaviour.* Oliver & Boyd. London.

YANG, J. 1982. An estimate of fish biomass in the North Sea. *Journal Conseil International Exploration de Mer* 40: 161–172.

Index